the Song of the Earth

a synthesis of the scientific & spiritual worldviews

Worldview Key

Editors
Maddy Harland
and
William Keepin PhD

the Song of the Earth

a synthesis of the scientific & spiritual worldviews

Permanent Publications

Published by
Permanent Publications
Hyden House Ltd
The Sustainability Centre
East Meon
Hampshire GU32 1HR
United Kingdom
Tel: 01730 823 311
Fax: 01730 823 322
Overseas: (international code +44 - 1730)
info@permanentpublications.co.uk
www.permanentpublications.co.uk

Distributed in the USA by
Chelsea Green Publishing Company
PO Box 428, White River Junction, VT 05001
www.chelseagreen.com

First edition © 2012 Gaia Education & Permanent Publications
Reprinted 2013

Edited by Maddy Harland and William Keepin

Printed in the UK by
CPI Antony Rowe, Chippenham, Wiltshire

Printed on paper from mixed sources certified by the
Forest Stewardship Council

The Forest Stewardship Council (FSC) is a non-profit international organisation established to promote the responsible management of the world's forests. Products carrying the FSC label are independently certified to assure consumers that they come from forests that are managed to meet the social, economic and ecological needs of present and future generations.

British Library Cataloguing-in-Publication Data
A catalogue record for this book is available from the British Library

ISBN 978 1 85623 095 7

All rights reserved. No part of this publication may be reproduced, stored in a retrieval system, rebound or transmitted in any form or by any means, electronic, mechanical, photocopying, recording or otherwise, without the prior permission of Hyden House Limited.

Contents

Foreword	*Mark Richmond*	ix
Introduction	*Maddy Harland & William Keepin*	xi

1 The Holistic Worldview

Inner Net of the Heart: The Emerging Worldview of Oneness	*William Keepin*	2
Towards A Biomimicry Culture of Cooperation	*Elisabet Sahtouris*	18
A Gaian Worldview	*Ross Jackson*	25
The Dual Origin of the Universe	*May East*	36
Living the New Worldview: Global Justice & Saving Three Billion Years of Evolution	*Hildur Jackson*	41
Human Being as Miniature Galaxy	*William Keepin*	47

2 The Awakening & Transformation of Consciousness

Earth as Sacred Community	*Thomas Berry*	52
The Urgent Need for Spiritual Awakening	*Jetsunma Tenzin Palmo*	58
Spiritual Responsibility at a Time of Global Crisis	*Llewellyn Vaughan-Lee*	63
Living in Auroville: a Laboratory of Evolution	*MARTI*	73
Who Am I? Why Am I Here? Living the New Worldview	*Hildur Jackson*	81
The Great Transformation: The Opportunity	*David Korten*	91
The Great Turning	*Joanna Macy & Chris Johnstone*	93
The Shambhala Warrior	*The Venerable Dugu Choegyal Rinpoche*	99

3 Reconnecting With Nature

Ancient Prophecies and The Vision Quest as a Path to Oneness	*Hanne Marstrand Strong*	102
The Declaration of the Sacred Earth Gathering	*The Wisdom Keepers*	108
Riding the Paradox: A Colourful Middle Way	*Pracha Hutanuwatr & Jane Rasbash*	110
The Bioregional Vision	*Gene Marshall*	116
Voices of our Ancestors	*Dhyani Ywahoo*	118
Pathways to Integration: Rediscovering the Song of the Earth	*Maddy Harland*	120
Seeding the Round Planet	*Stephan Harding*	127
Japanese Haiku	*MARTI*	129

4 Health & Healing

The World as a Holowave: Theory of Global Healing	*Dieter Duhm*	134
Planetary Healing: A New Narrative	*Maddy Harland*	142
Healing Ourselves	*Maddy Harland*	148
The Power of Reconciliation and Forgiveness	*Duane Elgin*	155
The Intelligent Heart	*Michael Stubberup & Matias Ignatius*	163
Peace Circle Dialogues: I Am Because You Are	*Karambu Ringera*	170
The Cracked Mirror	*Wangari Maathai*	174
Maher – Rising to New Life: Interview with Lucy Kurien	*William Keepin*	179
Health in the Global South	*Rashmi Mayur*	185
A Healthy Lifestyle	*Dr Cornelia Featherstone*	187
The Dream of the Children	*Sabine Lichtenfels*	190

5 Socially Engaged Spirituality

InterSpirituality: Bridging the Religious and Spiritual Traditions of the World	*William Keepin*	194
The Spiritual Imperative	*Satish Kumar*	202
Guidelines for Socially Engaged Spirituality	*William Keepin*	214
Silence and the Sacred: Interview with Craig Gibsone and Robin Alfred	*William Keepin*	220
Spirituality in Damanhur	*Macaco Tamerice*	235
A Brief Snapshot of Auroville, India	*MARTI*	242
Plum Village: A Spiritual Perspective on Community	*MARTI*	244
The Awakening of the World, the Village, the Nation and the World: The Sarvodaya Vision for the Global Future	*Hildur Jackson*	246
A Hopi Elder Speaks	*The Elders Oraibi*	254

The Editors

Maddy Harland co-founded Permanent Publications with her husband Tim in 1990. They launched *Permaculture* magazine in 1992 to promote small-scale, positive, low carbon solutions. Maddy became the editor in 1994 and the magazine is now read in 77 countries by more than 100,000 people. The company has won many awards for its work, notably a Queen's Award for Enterprise in 2008 for 'its unfettered dedication to promoting sustainable development internationally'. Maddy also co-founded The Sustainability Centre in 1995, a former Naval base, which is now a thriving educational charity on the South Downs in Hampshire, England. Maddy is a Fellow of the Royal Society of Arts and an accredited teacher with the International Network of Esoteric Healing. She lives on the South Downs and has two daughters. When not losing herself in her permaculture garden, her writing can be most regularly be found on **www.permaculture.co.uk**

William Keepin, PhD, is co-founder of the Satyana Institute and founder of the Gender Reconciliation International project which leads intensive training programs in several countries for healing and reconciliation between women and men (**www.GRworld.org**). A mathematical physicist with thirty scientific publications on sustainable energy and global warming, he became a whistleblower in nuclear science policy (recounted in *The Cultural Creatives* by Ray and Anderson). He has trained extensively in spiritual traditions East and West, and has facilitated Holotropic Breathwork for 25 years (Grof Transpersonal Training). William is a Fellow of the Findhorn Foundation and adjunct faculty at Holy Names University. He leads contemplative retreats on the emerging interspirituality that bridges major world religions and science (www.pathofdivinelove.org). His previous books are *Divine Duality: The Power of Reconciliation Between Women and Men* (2007) and (with co-author Cynthia Brix) *Women Healing Women* (2009).

Foreword

Under the Patronage of

United Nations Educational, Scientific and Cultural Organization

United Nations Decade of Education for Sustainable Development 2005 – 2014

The Song of the Earth is the fourth book in the series of 'Four Keys to Sustainability' and supports the Ecovillage Design Education (EDE) curriculum designed by Gaia Education. Gaia Education has been an active partner of the UN Decade of Education for Sustainable Development (DESD, 2005-2014), since its launch in 2005. The first three Keys – Social, Economic and Ecological – explore the practical means by which humanity can learn to live together in sustainable, life-enhancing ways: conserving biodiversity, promoting Earth restoration and ecological balance, social harmony, global financial stability and a balance of resource consumption and wealth between the industrial and developing nations. These are explicit, tried and tested ideas, technologies and techniques that ecovillages all over the world have been successfully experimenting with for many decades.

The Fourth Key, *The Song of the Earth*, is more implicit. It explores an emerging worldview which is shared between cultures all over the planet. It celebrates the ageless insight that humanity is part of the interconnected web of life in a dynamic planetary system. The book asks important questions such as: how does this understanding shape our thoughts and subsequent decisions and actions? How can we not only survive but prosper together whilst protecting the planet's finite resources for future generations? How do we create a fairer, more equitable world that is capable of generating peace?

These key questions are explored through the voices of many authors from all parts of the world. What is shared between all the contributions is the insight that systemic, integrated ways of thinking will naturally result in more sustainable ways of living. Once our deep interconnection with our fellow human beings, other species and our living planet is a reality, it should no longer be possible to damage the web of life.

From this arises an understanding of the unity between all peoples and our dependence on our home, the Earth. This is an important step for humanity: we are no longer 'in dominion' but now part of a whole planetary system. As Thomas Berry, the theologian said, "We have to reinvent ourselves at a species level." Yet there is evidence that this big step has slowly yet progressively been taken since the last century all over the world.

The United Nations has been an early pioneer of this work since 1945, bringing together humanity to promote human rights and world peace. The UN Decade of Education for Sustainable Development pursues this vision by seeking to integrate a fundamental education concept into all areas of education and learning: every human being should benefit from an education that provides the values, competence, knowledge and skills required to shape the future in line with the demands of sustainable development.

It is therefore exciting to see a convergence of integrative thinking, and note the many practical examples of how these values already shape sustainable communities in all continents and help create active and responsible citizens ready to address global sustainability challenges and to act in an increasingly complex world. This is an insightful book that carries a message full of hope.

MARK RICHMOND
Director, Division of Education for Peace and Sustainable Development
UNESCO

Introduction

Hildur Jackson originally conceived the idea of a book that would gather together the wisdom of the integral worldview from different continents. It was to be the core that held together the three-legged stool of ecological, economic and social sustainability. How we perceive the world shapes our relationships and behaviours, and to truly practice sustainability, we need to hold an integral worldview at the centre of our being. From the very onset, this book aimed to explore what this might mean, not only as a set of philosophical ideas but as a living experiment. Thus, as we collected stories, interviews, articles and ideas from all over the world, we would create a pattern of this integral worldview for others to build upon.

We decided to call this book *The Song of the Earth* to celebrate our beautiful Mother Earth, alive in a vast cosmos, and thereby acknowledge a new emerging note that is being sounded all over the planet. As our post-industrial civilisations begin to crumble, we are witnessing the limits of our natural resources and the inability of our economic and social systems to sustain themselves. At the same time there is a cogent global shift amongst the peoples of the world who are seeking new ways of living less materialistic, more connected lives.

People all over the globe yearn for what they know in their hearts is possible: a new human civilization of harmony and cooperation between all peoples – dwelling in balance with the Earth. This vision is not a dreamy fantasy; it is the birthright of humanity. A new and powerful worldview is rapidly emerging today in response to this deep yearning, accompanied by a growing spectrum of 'new-paradigm' literature and websites. Because the issues are inherently vast and complex, most writers necessarily focus on one or another aspect of the larger inspiring shift that is taking place. Most books on the subject emphasise one aspect or another – either the spiritual or the ecological or the political or psychological or the North/South dimensions of a much larger cultural transformation that is taking place today. Much of this literature analyses and advocates the changes that are needed – and rightly so – but much less attention is given to exploring the steps of how these changes are to be carried out in practice.

By contrast, the literature emerging from the diverse and loosely-affiliated international array of spiritual, ecological, and intentional communities –

often collectively dubbed the Ecovillage (EV) Movement – provides a framework for a truly integral worldview that weaves in many practical details and time-tested strategies. Taken together, these ecovillage projects comprise a largely unprecedented global experiment in sustainable living for this century and beyond. The resulting worldview emanating from this global experiment is as practical as it is inspiring, because it emerges from diverse projects and people who are actually LIVING the new civilization today, rather than merely advocating for it in some distant or hoped-for future.

This background provides the motivation for this series of four books we are calling the Four Keys. Taken as a whole, the ecovillage movement is bringing forth comprehensive lessons and practical findings from a comprehensive 40-year cross-cultural experiment in sustainable living. And the results are remarkable: a community settlement in northern Scotland that has the lowest ecological footprint anywhere in Europe, while maintaining a high quality of life; communities of women and men of diverse religious faiths living together in unprecedented harmony in the midst of the highly patriarchal and religiously-divided India; a suburban ecological community in upstate New York that shares resources, grows food, even has its own currency.

Indeed, the larger ecovillage movement is achieving sustainable community in human, ecological, and spiritual terms in ways undreamt of even by some of our most forward thinking visionaries. How many environmental firms and organizations can claim to produce zero-waste effluents from their own offices and buildings? How many spiritual communities are self-sufficient in terms of energy consumption, consuming a net of zero from the electric power grid, thereby contributing virtually nothing to global warming? How many psychologists, therapists, and social workers are forging new modalities of healing and transformation in community, rather than simply treating clients to adjust themselves to an inherently alienated, materialistic society? How many religious leaders are cultivating authentic spiritual awakening and striving to bridge destructive religious divisions in their communities, rather than promulgating tired doctrines and stale rituals that keep their congregations asleep?

Of course the noble aspirations and vital contributions of all people who strive for a whole and transformed society are profoundly needed. We advocate and celebrate the innumerable effective, inspirational and practical solutions emerging today to the world's problems. We need them all. What makes the ecovillage movement so unique, however, is its practical and sustained experimentation in synthesizing the major aspects of an ecologically sound, spiritually grounded, nourishing human society. Taken together, the ecovillages across the globe today are pioneering an unprecedented global community experiment – a living laboratory for birthing a new humanity. The worldview emerging from this larger movement is not mere philosophy. It is the new paradigm – alive and actively manifesting itself fully and beautifully on the ground – like a fractal seed, in miniature, of the coming civilization.

What does this mean to all people, however, and not just the fortunate few who are living in ecovillages, pioneering these new paradigms? Years ago Maddy was deeply affected by a story Joanna Macy told her about the rainforest activist, John Seed. Joanna described him standing in front of a hostile crowd of loggers with bulldozers in a rainforest. Suddenly, John had an epiphany. He realized that he was not just one small individual protester. He was the rainforest itself. He was the Lifeforce of this beautiful, biodiverse and irreplaceable sanctuary, and he was the intimate web of life that flowed through the landscape. The rainforest took the shape of his body; it flowed through him, within him, throughout him. He could 'think like a mountain' and that force transformed his physical body as well as the power of his intentions. He later wrote, "There and then I was gripped with an intense, profound realization of the depth of the bonds that connect us to the Earth, how deep are our feelings for these connections. I knew then that I was no longer acting on behalf of myself or my human ideas, but on behalf of the Earth… on behalf of my larger self, that I was literally part of the rainforest defending herself."[1]

This is the essence of deep ecology, to realise our oneness with the web of life – not just as an aspect of systems science or as an understanding of applied ecology – but as authentic knowledge, consciousness. This is part of the integrated worldview, but it is not the only story. There is more.

Whilst Maddy was working on this book she had a powerful experience, a waking dream, which opened her understanding of the emerging worldview. Deep in the night she dreamt that her consciousness had merged with the Lifeforce of all things. She was for a time transported to an inifinite dimension that held all that she knew in miniscule. This was a vast level of being, so different that it holds no reference to our temporal world. For that time she had direct experience of the interconnected Oneness of all things. She knew that in this place of awareness all actions would flow from a centre that was completely at peace, suffused with love, fearless yet harmless to all beings. There was no need of 'morals' or striving for 'right action', as these values arise naturally from unified consciousness. This is the deep secret of peace. We cannot commit self-harm or harm others when we are in love with Life; we cannot damage the web of Life when we are the web and the Life itself. From our consciousness our actions arise.

This experience was deeply powerful as it demonstrated in an unforgettable way that whilst we may design our lives sustainably using technologies, knowledge and practices drawn from best practices within the fields of ecology, conservation, agriculture, Gaian economics, social justice and wise governance… unless we are able to draw upon an integral worldview we are still pasting the pieces together in the old disconnected paradigm. We urgently need this unifying vision and experience of the inherent oneness of all Life. Otherwise we are like children attempting a complex jigsaw without a picture. We lack an orchestrated synthesis, and so we have yet to respond from a place that holds an awareness of the Whole.

The Song of the Earth is our attempt to create such an integral picture. It is our call to explore the marvellous web of life on our planet, in our cosmos, and to take those insights 'home', *oikos*. It asks that we take a step in human evolution and begin the work from awareness as if we are all an interconnected One. It asks that we consciously design gentle, benign, sustainable lifestyles and communities that honour diversity in all forms: social, cultural, spiritual, as well as animal, plant and mineral. This is the 'Song of the Earth' and to listen to its exquisite music has never been more urgent. Rather than being impelled by exhausted resources, limits to growth and human-induced crises, let us instead be inspired by the possibilities of the new world and new Life that is emerging rapidly now.

MADDY HARLAND AND WILLIAM KEEPIN
Editors

MODULE 1
The Holistic Worldview

Contents

Inner Net of the Heart:
The Emerging Worldview of Oneness

Towards a Biomimicry Culture of Cooperation

A Gaian Worldview

The Dual Origin of the Universe

Living the New Worldview

Human Being as Miniature Galaxy

History reveals that when humanity is faced with new challenges that cannot be solved with old thinking, new capacities at mental and biological levels will evolve. We are now living in a point in history when changing life conditions are of such magnitude that a new worldview with a transformative vision is beginning to emerge.

Nancy Roof, PhD, co-founder, United Nations Values Caucus

This chapter articulates a new worldview that is rapidly emerging and encompasses a fresh, unifying perspective on human existence, science, spirituality, and humanity's place in the larger web of life.

Inner Net of the Heart: The Emerging Worldview of Oneness

William Keepin, PhD

At the cutting edge of science today, a powerful new awakening is taking place in many different disciplines. The key insight is this: beyond the physical realm, there exist invisible patterns and principles that somehow organize what we observe and experience in the physical world. Science is discovering that 'something transpires behind that which appears'. This startling theme is emerging in field after field: including biology, physics, non linear dynamics, artificial life, brain physiology, complexity theory, transpersonal psychology, psycho-neuroimmunology, and ethnobotany – to name a few. This auspicious development points Western science in a remarkable direction toward the existence of a realm beyond the observable, material, empirical world. The stage is steadily being set for another major scientific revolution – one that will eventually weave science and spirituality – matter and spirit – together into a seamless unity.

At the same time, a parallel awakening is taking place across the world's spiritual and religious traditions. As East meets West and North meets South, the oneness of all life is being revealed and manifest in unprecedented ways, and the essential unity of human spiritual consciousness is becoming increasingly recognized. The world's wisdom traditions and religious disciplines are linking together, bringing a convergence and collaboration among previously disparate practices and teachings.

What is the meaning of these trends, and where are they headed? We are witnessing nothing less than a revolution of consciousness – the birth of a vast, integral worldview that unites and cross-fertilizes East and West, modern and indigenous, human and non-human, contemporary and ancient

– all leading us toward a deep collective realization of the seamless oneness of existence. And this auspicious breakthrough is coming not a moment too late, because humanity urgently needs a comprehensive, unifying perspective in order to navigate our way through the challenging waters ahead, and begin building a new civilization of love and harmony. This is not a dreamy mystical vision. It is our destiny as the human family, and it is our birthright to bring this destiny into concrete manifestation.

This chapter articulates this new emerging worldview, which is sometimes called holistic, or integral, or the new paradigm; there are many names for it. We begin by examining the trends mentioned above more deeply; specifically the growing links between science and spirituality, and the remarkable correspondence between certain recent scientific discoveries and the ancient wisdom of mystics and sages down through the ages. Next we consider the emerging commonalities among the world's spiritual and religious traditions, and how certain universal truths seem to link them together. Finally we consider the larger implications of these epochal shifts and key features of the auspicious new worldview they portend.

The Emerging Synthesis of Science and Spirituality

The traditional Western scientific worldview has conditioned us to believe that the world and the cosmos are comprised of distinct, isolated, material objects – all separated from one another in space and set in dynamic motion according to rational, deterministic, mechanistic laws. This view of the universe reigned supreme for hundreds of years or more, bolstered by the observations of classical physics, biology, and other natural sciences. However, in the past century it began to crumble, driven by numerous remarkable discoveries at the frontiers of science. This led to a fundamental shift in perspective that began about 100 years ago with the advent of quantum theory and relativity theory in physics, and continued through the 20th century with major breakthroughs in biology, evolutionary biology, complexity theory, transpersonal psychology, and many other disciplines.

A new scientific paradigm is emerging that conceives of the universe as a vast network of energetic living systems, all interconnected in a complex web of relationships that manifest out of an underlying unified field. Atoms, once believed to be solid nuggets of matter, are revealed to be patterns of vibrating energy. Matter is composed almost entirely of empty space. A new holistic or integral worldview is emerging, in which the universe and all life in it comprise a single unitive whole.

These unprecedented discoveries are rapidly shifting our understanding of reality. Science is uncovering profound new levels of interconnection between matter and consciousness, which began with Heisenberg's discovery that nothing exists in objective isolation from the rest of existence. For example, experiments in quantum physics show that every particle in the universe is somehow 'aware' of every other particle. A new principle of interconnection known as 'non-locality' has been confirmed in repeated experiments, showing that after two particles have interacted, their 'spin' properties are thereafter fundamentally interconnected. If we measure the spin state of one particle, the spin state of the other particle responds instantaneously, even if it's located at the opposite end of the universe. This immediate 'non-local' connection between the particles transcends all separation in space and time, and thus precludes any possibility of a 'signal' of some kind carrying information from one particle to the other. Thus the entire universe appears to be intimately interconnected, and these discoveries hint at possible physical explanations for the increasingly recognized links between consciousness and matter.

Consciousness Research at the Frontiers of Science

Remarkable new discoveries are pointing to previously unknown links between consciousness and matter. A body of striking evidence has been generated recently through the Global Consciousness Project. Sensitive electronic detectors called random number generators have been placed in 40 countries across the globe and monitored closely around the clock for several years. Most of the time, these detectors produce a steady stream of random data. What is remarkable, however, is that whenever major world events occur – times when the consciousness of millions of people is focused on the same event at the same time – the detectors deviate radically from their usual data patterns. In this manner, the detectors have sensed a whole range of major world events as they were happening, such as Princess Diana's funeral, or the hung U.S. Presidential election in 2000. What is most remarkable is that the detectors seem to sense major events even in advance of their occurrence. For example, not only did the detectors strongly register the terrorist attacks of September 11, 2001, but the tell-tale shift in the data pattern began four hours before the two planes struck the Twin Towers. Similarly, in December, 2004, the detectors began registering the devastating tsunami in south east Asia a full 24 hours before the massive underwater earthquake took place.

In a similar vein, Princeton University scientist Robert Jahn and several colleagues at the Princeton Engineering Anomalies Research laboratory conducted 27 years of rigorous experiments at the frontiers of consciousness and science. They found that human wilful intention could skew the otherwise random behavior of sensitive electronic and physical machines in a volitional direction. Although the effect was weak (one part in 10,000), the results were highly statistically significant, with the odds of occurring by chance only one in a trillion. The same researchers also conducted 653 experiments in 'remote viewing', a controlled form of what is popularly called clairvoyance, and they concluded that information was reliably transferred between human beings telepathically. Again, the probability of these results happening by chance was minuscule – one in 33 million. Even more strikingly, the effects were shown to be independent of the spatial separation of the 'sending' and 'receiving' agents, as well as independent of temporal separation up to several days. These startling results seem to replicate on a macroscopic scale the 'non local' interconnection discussed above for elementary particles on a microscopic scale.

In his recent book *Entangled Minds*, scientist Dean Radin compiles the results of similar research experiments emerging from laboratories all around the world. Taken together, Radin maintains that scientific research 'has resolved a century of skeptical doubts through thousands of replicated laboratory studies' that demonstrate the reality of phenomena traditionally considered impossible by today's mainstream science, such as extrasensory perception (ESP), psychokinesis, and telepathy. For Radin, these developments point toward a whole new awakening of the role of consciousness in physical reality. And such findings accord with what many people already believe, based on personal experience. According to a 2005 Gallup poll in the United States, about 75% of American adults believe in at least one such 'paranormal' phenomenon, including 41 percent who believe in ESP, and 55% percent who believe in the power of the mind to heal the body. In most cases, people come to these beliefs through personal experience or anecdotes, rather than through studying the scientific evidence.

On a more practical human level, extensive research has been conducted on the healing effects of prayer and meditation. Psychiatrist Elisabeth Targ conducted a randomized, double-blind study on the effects of prayer for patients with AIDS, published in the *Western Journal of Medicine*. She found that subjects who were prayed for had significantly better health and lower morbidity than the subjects who were not prayed for. Physician Larry Dossey has explored this phenomenon in depth, reviewing hundreds of scientific research studies on the efficacy of prayer. He concludes that focused intercessory prayer improves healing of medical patients. He further found that this healing effect was independent of the spatial separation between the patients and those who were praying for their healing. To explore whether a 'placebo' effect of some kind might be a possible explanation for these data, Dossey conducted several experiments on the connection between human consciousness and replication rates of seed germination and bacteria growth in the laboratory. He found that the biological replication rates can be directly affected by human wilful intention. Because the bacteria could not have been aware that they were being 'prayed' for, Dossey concludes that there must exist a fundamental interconnection between intercessory prayer and biological and physiological processes. Compiling findings from hundreds of scientific experiments on prayer and meditation, Dossey concludes that the evidence is 'simply overwhelming that prayer functions at a distance to change physical processes in a variety of organisms, from bacteria to humans'.

Debate continues over such findings, and these unusual phenomena remain difficult for many mainstream scientists to swallow, even when well supported experimentally. Yet this may simply be the result of a widespread lack of awareness about the evidence. Dean Radin maintains that owing to 'a general uneasiness about parapsychology' and the 'insular nature of

> *All things by immortal power,*
> *Near and Far*
> *Hiddenly*
> *To each other linked are,*
> *That thou canst not stir a flower*
> *Without troubling of a star.*
> Francis Thompson, 19th century poet

scientific disciplines, the vast majority of these experiments are unknown to most scientists'. Yet Radin is convinced that anyone who makes a serious and honest inquiry into the entire body of research would have to agree that something real is going on that cannot be explained by today's science.

These discoveries show that matter and consciousness are connected in profound ways we don't yet understand. They also point toward a remarkable convergence between the new scientific understanding and spiritual wisdom through the ages. It appears that science is just now discovering what mystics have affirmed down through the ages; matter and consciousness are fundamentally interconnected, and all things in the universe are part of one, integral whole.

Uniting Spirit and Matter: Beyond $E = mc^2$

To take a closer look at what is emerging at the frontiers of science, let us examine one specific discipline more closely to see the essential shift that is taking place across so many disciplines. We will choose physics, the 'hardest' of the sciences. The reader may rest assured that no knowledge of physics is required to follow this section – only an eye for beauty.

We begin with the work of physicist David Bohm (1917-1992), a leading pioneer in science and spirituality. Bohm was a colleague of Einstein's at Princeton, and the two of them shared similar views on the theoretical foundations of quantum theory. Bohm was driven by a deep passion to understand the nature of the Universe. He felt the true purpose of science was a quest for truth, and it disturbed him that many scientists regarded science primarily as a pragmatic means for prediction and control of nature and technology. Like Einstein, Bohm believed that science was a kind of spiritual quest, a deep search for truth or *jnana* yoga that strives to discover the ultimate secrets of existence.

Bohm probed deeply not only into modern physics, where he made major contributions, but he also carried his quest well beyond science itself. Bohm delved deeply into spiritual teachings and wisdom, and for more than 20 years he carried on in-depth dialogues with the Indian sage, Krishnamurti, and other leading spiritual masters including the Dalai Lama. He also explored art, to discover insights about the nature of order and form. Bohm eagerly embraced both scientific and spiritual forms of inquiry as a way to 'triangulate', so to speak, on the true nature of reality by taking into account the broadest possible range of data and methods of inquiry.

Bohm began by asking what the twin pillars of the new physics – relativity theory and quantum theory – had in common, and he discovered that it is wholeness. Both theories proposed that the universe is a single integral whole, from the tiniest atoms to the largest galaxies. Building upon this foundation over 30 years of rigorous scientific work, Bohm emerged with the hypothesis that the essence of the universe is what he called the holomovement. 'Movement' means that the nature of existence is a process of continual change, and 'holo' means that it has a kind of holographic structure, in which each part contains the whole. In Bohm's words, "The cosmos is a single, unbroken wholeness in flowing movement," in which each part of the flow contains the entire flow.

The Implicate Order

Bohm further proposed that there are two fundamental aspects to the holomovement, which he called the explicate order and implicate order. Now one might well ask, after having just said it is oneness, why are we breaking it into two? Aren't we imposing a false duality onto what is a unity? Not quite, because the explicate and implicate orders only appear to be distinct; in truth they are unified. But they appear convincingly separate because of human perceptual limitations. We humans have five physical senses plus the thinking mind, which together perceive only a small portion of the oneness. This limited portion – that which is directly perceived by our six human faculties – constitutes what Bohm calls the explicate order. Everything else – all that we don't directly see, hear, taste, feel, touch, or think – constitutes the implicate order. In summary, then, the limitations of human perception necessitate a delineation between what is directly perceptible to the senses (explicate order), and what isn't (implicate order).

To illustrate the relationship between the implicate and explicate orders, Bohm gave a simple example. Imagine two concentric cylinders, one larger than the other, and suppose the annular column between them is filled with a thick transparent liquid like glycerin. Now place a small droplet of ink on the top surface of the glycerine, and begin rotating the inner cylinder (while the outer cylinder remains fixed). As the rotation continues, the ink droplet gets stretched out and becomes longer and thinner, growing ever fainter. Eventually, it disappears altogether. At this point, the natural conclusion to draw is that the order, or organization, of the original ink drop has been lost – rendered chaotic – as the ink seems to be randomly distributed in microscopically small particles throughout the glycerin. However, if you now rotate the inner cylinder in the opposite direction, the ink structure will begin to reappear, faintly at first, and as you keep rotating it gets thicker and darker, until it finally comes all the way back; the ink droplet reconstructs itself completely. The key insight from this experiment (which has been demonstrated in the laboratory) is that the order in the original ink drop was preserved – enfolded in the glycerine – even when it was no longer visible.

Bohm used this example as a metaphor to emphasize a fundamental lesson for all of science: "A hidden order may be present in what appears to be random." This seemingly simple, almost self-evident statement has profound implications. It is a scientific statement of the famous quip in Shakespeare's *Hamlet*, "There are more things in Heaven and Earth, Horatio, than are dreamt of in your philosophies!"

Developing the full implications of his theoretical work, Bohm proposed that there exists a vast, invisible realm beyond what we perceive as the physical universe – yet which is every bit as real as the physical universe.

He called this hidden realm the 'implicate order', and he demonstrated that the implicate order is not only consistent with the data of modern physics, but in fact it provides a cogent and comprehensive explanation for all the seemingly bizarre phenomena of quantum and relativistic physics.

At first blush, it's natural to suppose that the implicate order is some kind of secondary, ethereal reality – floating around somewhere in space, whereas the primary reality is the solid physical universe, just as our senses perceive and our science describes. Yet for Bohm, precisely the opposite is the case. The implicate order is the fundamental reality, and the explicate order is a secondary phenomenon. The explicate order is analogous to the foam on the waves of the ocean, and the implicate order is the ocean itself. Sea foam is certainly lively, beautiful, and vast in its own right – extending across the entire planet. But compared to the ocean, sea foam is but a minor and ephemeral 'surface phenomenon'. Just so, the explicate order – the physical universe with its 100 billion galaxies, each of which contains 100 billion stars – is a kind of ephemeral side effect or by-product, created by the far vaster implicate order. This does not in any way diminish the reality, beauty, or sacredness of the physical universe, but simply places it in proper relationship to the unmanifest dimension. It is reminiscent of what God (Krishna) tells the warrior Arjuna in the Bhagavad Gita: "I run this entire cosmos with the tiniest part of my Being." The implicate order is profoundly vast – a kind of interpenetrating field of conscious presence and intelligence that far transcends the physical universe, yet which creates the universe.

What is the nature of the implicate order? It is present everywhere, but visible nowhere. It extends throughout space and time, but also far beyond space and time. This is a crucial point to understand; space is not some giant vacuum through which matter moves. Rather, matter and empty space are intimately interconnected, and both are part of the explicate order. The implicate order transcends space and time altogether, although it interpenetrates with every point in space-time. One can think of the implicate order as a synonym for the Unseen, for that which is neither manifest nor accessible to our mind and five senses – often called the spiritual dimension. We don't directly perceive it, except through inner intuitions and contemplative forms of spiritual practice.

Matter, Energy, and Consciousness

A vital aspect of Bohm's thinking is that the nature of reality has three fundamental components. Science has generally dealt with only two of them: matter and energy. These two are equated in the famous equation from Einstein: $E = mc^2$. This equation essentially affirms that energy and matter are different forms of the same thing. Bohm insists there is a third element, which he called 'meaning' or consciousness (he equated the two terms). For Bohm, consciousness is at least as significant as matter and energy, and he proposed a tripartite structure to reality: matter, energy, and consciousness. Moreover, each of these basic building blocks contains or 'enfolds' the other two. Thus, energy enfolds both matter and consciousness. Similarly, matter enfolds energy and consciousness, and consciousness enfolds both matter and energy. Bohm reaches a powerful conclusion: "This implies, in contrast to the usual view, that consciousness is an inherent and essential part of our overall reality, and is not merely a purely abstract and ethereal quality having its existence only in the mind."

Consciousness encompasses the invisible aspects of life – purpose, yearning, intention, love, despair, all of the intangibles of life – which are no less real for being intangible. They are just as real as matter or energy, but they cannot be measured in the scientific laboratory. In fact, scientific instruments can be viewed as simply technological extensions of our five physical sense perceptions. Microscopes and telescopes are just bigger eyes. Microphones are high-tech, mechanical ears. Bohm emphasizes that these and all scientific instruments operate only in the explicate order, and they register just a small fraction of reality. Conventional science therefore misses the implicate order altogether, and with it, the entire domain of consciousness. The frontiers of the new science are finally beginning to open to this domain.

... consciousness is an inherent and essential part of our overall reality, and is not merely a purely abstract and ethereal quality having its existence only in the mind.
David Bohm

As Above, So Below

The final and perhaps most important characteristic of Bohm's new scientific theory of reality is its holographic structure. To illustrate what this means, let us consider an example from mathematical physics – fractal geometry – called the Mandelbrot set. Over the past 30 years, new mathematical structures have been discovered that are very useful for scientific modelling of a broad range of natural phenomena. These structures are called fractals, and the simplest example is the 'Mandelbrot' set shown in Figure 1 (see overleaf).

Who Are You? A Simple Inquiry

To illustrate the key insights here, let us engage in a simple imagination exercise. Imagine for a moment that the figure shown in Figure 1 is a model of the entire cosmos, a kind of photograph or snapshot of everything that exists. Somewhere deep inside this figure is you – a single human being, sitting there as a tiny spec in the vast universe. Of course, you are too small to see. So let us now zoom in on Figure 1 to find you; we want to get a good look at you, and your relationship to the cosmos.

We begin by magnifying the portion near the left hand side of Figure 1. This gives us Figure 2, and we keep zooming in further. Magnifying the center portion of Figure 2 gives us Figure 3. Then we zoom in on the center of Figure 3, which gives us Figure 4, and so on. Continuing in this fashion, we magnify the center of each figure to generate the next one.

Because your form is small compared to the whole cosmos shown in Figure 1, we don't even see a hint of you until we get to Figure 6. There,

The Mandelbrot Set.

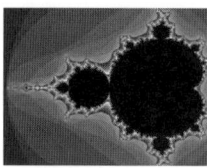
1

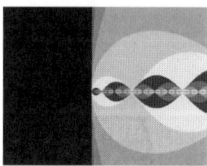
2

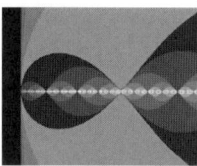
3

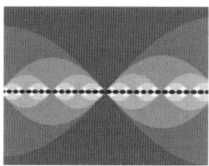
4

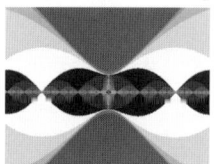
5

6

7

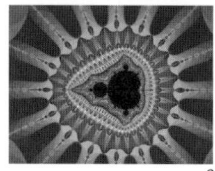
8

we finally catch a glimpse of you at the very center of the star shape in the middle of Fig. 6. But we still can't see you very well, so let's zoom in again. In Figure 7, we can just begin to see you – a small black dot in the center. So we zoom in one more time to get a better look at you, and in Figure 8, there you are! We finally see you clearly.†

And here in Figure 8, we discover something remarkable: you are an exact replica of the first figure! You are a miniature version of the entire cosmos. This is a breakthrough realization, a major "Aha!" You realize that your true nature is absolutely identical to the true nature of the entire cosmos. You and the cosmos are one!

How can this be? What does it mean? All the richness and mystery and depth of the cosmos – the Infinite – is right there inside you. As the mystical poet Rumi describes this realization, "The secret turning in your heart is the entire Universe turning!"

The remarkable mathematical structure shown in these figures gives a metaphorical representation of the holographic nature of life and consciousness. In science, this phenomenon is described as 'nested sets of self-similar structures'. Although new to science, this principle has been known to sages and mystics since ancient times: "As above, so below." The fractal is thus a modern scientific discovery of an ancient alchemical principle, which can also be stated: "As within, so without." The key insight here is that deeply embedded within universal structures are miniature replicas of the whole, on vastly smaller scales. The microcosm replicates the macrocosm.

Of course on the physical plane – the explicate order – your form is indeed just a tiny spec. So your outer form is a spec, but your inner essence is the vast cosmos itself. And this is true not only for you, but for every other being as well. Our hearts are bigger than the universe.

The doorway to this universal consciousness is through our heart which opens inwardly to the implicate order that links us all together. This vast inner oneness could be called the 'inner net of the heart'. The analogy to the 'internet' here is intentional, because the computerized internet may be seen as a technological manifestation of this fractal principle in the explicate order. Every computer has access to the entirety of information on the internet (apart from electronic firewalls that cordon off domains of cyber security). Any part of this vast cyberspace is only a few clicks away. Indeed the very existence of the internet is a result, in the explicate order, of a pre-existent and far more refined parallel principle in the implicate order. Just as every computer can access the entire universe of cyberspace through the internet, every human heart can access the entire cosmos of consciousness through the inner net of the heart.

Glimpsing the Unity of Science and Religion

What are the implications of this? Although this fractal model is only metaphorical, it reflects powerful insights when applied beyond science proper, to various spiritual and religious traditions. For example, in Hinduism, the 'Atman' represents the spiritual nature of the individual (sometimes called the Self), and 'Brahman' is the spiritual nature of the cosmos. The fundamental enlightenment experience is the realization that Atman is Brahman – the two are identical. Similarly in the Bhagavad Gita, we 'see the Self in every creature, and all of creation in the Self'. In Judaism, "You are made in the image of God." These insights are precisely mirrored in the fractal imagery above: the tiny Mandelbrot set in Figure 8 contains all the richness and beauty of the large set in Figure 1. The tiny version suffers no loss of complexity or fine detail, even though it is more than 100 million times smaller. Another expression of this same principle in Buddhist and Hindu mythology is called Indra's net, in which the whole universe is imagined as a vast lattice of glistening jewels, each of which reflects all the others in its own facets. Thus each individual jewel reflects or contains within itself the entire universe. This structure is analogous to Aldous Huxley's concept of 'holons', utilized extensively in the work of Ken Wilber and others.

Bohm's holomovement can thus be seen as essentially a synthesis of two ancient spiritual principles: (1) the Buddhist teaching of impermanence – the notion that the nature of manifest existence is perpetual change (also put forth by Heraclitus), and (2) the microcosm reflects the macrocosm, as characterized for example in the Hindu mythological image of Indra's Net.

Similar parallels abound in other religions as well. In the Christian Gospels, "All that is mine is yours, and all that is yours is mine" (John 17:10), and "The Father and I are one" (John 10.30). The Christian mystic Julian of Norwich tells us, "We are all in God enclosed, and God is enclosed in us." Notice how these images are each a fractal-like expression of the identity of microcosm with macrocosm. Similarly in Islam, Allah says "Heaven and earth are too small to contain Me, but I fit easily inside the heart of my beloved devotee." In Zen, the great master Dogen says: "We study the self to forget the self, and when we forget the self, we become one with the ten thousand things." Here, the self we forget is just our physical and conditioned forms – our body, personality, ego, thoughts, family, vocation – all the attributes that characterize our manifest, temporal form. When we forget this self, we become one with the 'ten thousand things', meaning we become one with that which creates all of existence, i.e. the implicate order. Similarly, in the gnostic Gospel of Thomas, Jesus says: "When you make the two one, and when you make the inner as the outer and the outer as the inner and the above as the below, you enter the Kingdom." And finally, in Tantric Buddhism, the scholar Ajit Mukerjee says unequivocally: "The entire drama of the universe is replicated in the human body. When you come to know the truth of the body, you come to know the truth of the cosmos." And this is meant literally, but at a consciousness level, not a physical level. If you explore the nature of consciousness, you discover in your own being everything that goes on at the cosmic scale. As transpersonal psychologist Stanislav Grof emphasizes, "Each of us is everything."

> *The entire drama of the universe is replicated in the human body. When you come to know the truth of the body, you come to know the truth of the cosmos.*
> Ajit Mukerjee

In short: if you identify with your form and attributes, you are but a speck in the cosmos. But if you identify with your Being or essence, the whole of the Divine is merged into you, in all its depth and splendor. Your true identity is thus oneness with all that is – oneness with God. As Meister Eckhart put it, "I become all things, as God is, and I am one and the same being with Him... so entirely that 'He' and this 'I' become one 'is', and act in this 'isness' as one."

The Spiritual Unity of the World's Religions

The parallels quoted above between the fractal worldview in science and insights from diverse world religions also reflect an emerging fundamental unity between the religions. Mystics and sages from every religious tradition articulate a version of this fractal-like identity of the individual with the Divine (or God) – each using a different symbolic metaphor to express it. Indeed, all religions reflect this unity one way or another, because they all emanate from a single luminous core of spiritual truth. The world's multiplicity of religions are thus beginning to come together in a new way – to acknowledge a universal spirituality, and a fundamental unity of essential teachings.

The unitive insight is not new. Saints and sages down through the centuries have emphasized the essential unity of all religions. The Rig Veda put it succinctly thousands of years ago: "Truth is one. Sages call it by many names." The Sufi saint Al Halaj proclaimed the unity of all religions around 900 AD. The revered Hindu saint Ramakrishna proclaimed it again in the late 1800s, and during this same time period a whole new religion emerged in the Middle East that celebrates this essential unity (the Bahai faith).

In contemporary times, this trend is expressed in new ways, such as the 'perennial philosophy' articulated by Aldous Huxley, the Interspiritual Dialogue initiated by Wayne Teasdale and continued by Kurt Johnson, the integral spirituality of Ken Wilber, the Parliament of World Religions, and growing interfaith collaboration among spiritual and religious leaders across the globe. These auspicious developments are taking place despite the fact that widespread conflicts continue to be waged across the globe, constellated along lines of religious difference. There have always been political and fanatical elements within organized religion and politics that abuse and manipulate religious and spiritual teachings to justify persecution, hatred, conflict and war. The text of every major scripture can be twisted and distorted to lend seeming support to violence, plunder, and desecration. At the present time these destructive forces are sadly on the rise, particularly in the growing tension between the Judeo-Christian West and the Islamic nations.

Notwithstanding today's religious conflicts that are exceedingly destructive and will likely continue or increase in the short-term, the deeper underlying trend over the long term is in the opposite direction: a gradual but steady shift toward unification of the world religions, with a growing mutual respect and appreciation of religious diversity. Great strides forward were made in this regard during the 20th century, as the cultures of the world come into ever closer contact, and much greater strides will be made in the present century. The human spirit demands it, because the only path forward that will ever work in the long term is for the entire human community to live in harmony as one family and one species alongside billions of others species on this planet.

An important example of the emerging unity of religious teachings is the work of the Snowmass Conference, summarized briefly on pages 197-8 of this book. This group of spiritual leaders from nine major world religions has been meeting for more than 30 years, originally convened by the Christian monk, Thomas Keating. Each person in the group is a highly respected leader and mature practitioner within his or her respective faith. Over the course of their meetings, the Snowmass Conference developed eight points of common agreement. In effect, these eight points constitute an articulation of a universal spiritual faith – one that is consistent with the teachings of nine major world religions (Tibetan Buddhism, Theravadan Buddhism, Catholicism, Protestantism, Eastern Orthodox, Hinduism, Islam, Native American, Judaism). In the course of articulating these universal truths common to all their traditions, the group members also developed close interpersonal bonds and friendships with one another in the group.

While this accomplishment was encouraging in its own right, what is more striking about the Snowmass Conference is that after reaching major points of agreement, the group members then began to discuss their differences in religious beliefs and practices. They embarked upon this task somewhat hesitatingly at first, aware of major differences between their religions and not wishing to disturb the sense of unity and camaraderie they had already achieved. However, to their amazement and delight, what they discovered over time was that they bonded even more deeply over their differences than they had over their points of commonality. The richness and intricacies of their differences turned out to be fruitful ground for deep exploration together, and this process energized them and brought them even closer together as a group. Examples like this of true collaboration and friendship across religious differences are urgently needed in the world at this time of rampant religious conflict.

Religions in conflict with each other are like branches on a tree fighting with each other, not recognizing that they're all connected to the same trunk. The branches have their very existence only through that one trunk, which represents the mystical truth at the core of every religion. And the trunk just stands there, silently supporting and nourishing each branch, as they jostle around striving to win a trivial, unwinnable game.

The experience of the Snowmass Conference is an auspicious harbinger for future relations between the world's religions. A different and equally inspiring example is the Maher project described on pages 179-184. Here women from all different religious and social backgrounds live in harmony together and uphold universal love for their fellow humans – thereby creating a healing community that skilfully breaks the entrenched social taboos of India around religion, race, and caste that are some of the most divisive in the world.

When genuine spiritual leaders from different religious traditions come together, what emerges is rarely unbridgeable gulfs and conflicts, but rather rich and fertile ground that simultaneously (a) unites their respective teachings in a universal wisdom, and (b) honors and celebrates the uniqueness of each tradition. The plurality of religions is something to be cherished as a profound resource and gift to humanity – something that will be increasingly realized in the coming decades and centuries.

The emerging spiritual unity of the world's religions does not mean that the different traditions will fuse or unite into a single world religion. This is not the goal, nor is it desirable. Rather, each religion will take its proper place alongside the others, in mutual respect, to form a tapestry of traditions that together will embark upon an unprecedented level of spiritual work on behalf of humanity and the Earth. This process has already begun in earnest. According to Llewellyn Vaughan-Lee (see pages 63-72), there is now a certain inter-connecting work that can only be done by different spiritual traditions coming together in unique forms of collaboration and cross-fertilization.

A Sufi master, Pir Zia, gives a useful metaphor for understanding the synergistic relationship among the world's religions. He likens each religion to a physical organ in the body. Each organ – the heart, liver, brain, etc. – is unique, whole, and complete in itself, yet functions in harmonious concert with the other organs to sustain a living body. If any one organ of the body were to insist that the others do as it does, the body would die. If the heart, for example, insisted that the liver perform the same function as the heart, the body could not survive. Each particular organ is necessary, unique, alive, and complete in its own integrity, and it collaborates in synergy with the others to sustain life in the body. In analogous fashion, Zia posits that each world religion is an essential and necessary organ in the larger body of human spiritual consciousness. If any one religion imposes itself on the others, the healthy spiritual life of humanity is threatened. But working together, the diversity of world religions sustains a living, unified and vibrant body of human spiritual consciousness.

Toward a New Integral Worldview

Weaving these new developments and insights together, we are witnessing the dawning of a new science of consciousness and interconnection, or perhaps we might call it intercommunion. Thomas Berry summarizes the situation by affirming that the universe is not a collection of objects, but a communion of subjects – a beautiful synonym for what we have called here the inner net of the heart. Science is beginning to embrace consciousness itself as fundamental to reality. Laboratory experiments indicate that matter and consciousness are fundamentally interconnected, pointing toward an underlying unity that links all matter, energy, and consciousness. The oneness of all existence, long proclaimed by sages from many wisdom traditions, appears to be supported by the new science. Furthermore, consciousness is observed to transcend ordinary physical laws of matter, energy, space, and time. The physical universe appears to be dwarfed by the consciousness universe.

However, before mainstream science can fully embrace the rich promise of these new developments, it will have to loosen its grip on cherished doctrines of materialism and rationalism, which excessively restrict both its epistemology and ontology. Physicist Ravi Ravindra has observed that, "The greatest discovery of modern science is the discovery of its own limitations." Yet sadly today, many scientists still live as unwitting inmates in the conceptual prison of a narrow materialist worldview. Materialistic, mechanistic assumptions about the nature of reality limit many scientists' capacity to discover new forms of truth. The result is aptly summed up in Mark Twain's witty quip, "It ain't what you don't know that gets you into trouble, it's what you think you know that ain't so."

Simultaneous with these discoveries at the frontiers of science, the spiritual and religious traditions of the world are moving slowly yet inexorably closer together – toward a kind of universal spirituality. Leading voices from diverse spiritual and religious traditions are proclaiming the fundamental unity of wisdom from all the traditions.

Human consciousness is becoming widely recognized to be one with the consciousness of the universe. This applies to every human being, and every person has the potential to access this universal consciousness. This is not a dreamy, mystical metaphor – it is a literal truth of consciousness, and has been known by mystics for ages. As Rumi put it in the 13th century, "Let the drop of water that is you become a hundred mighty seas. But do not think that the drop alone becomes the Ocean. The Ocean, too, becomes the drop." It is every human being's birthright to discover this inner net of the heart within themselves, and to live from that vast interior foundation of consciousness.

A key consequence of this worldview is that if we are connected to and led by this larger wisdom of universal consciousness, then our actions and work in the world can become profoundly transformative. Human beings can be used by this larger wisdom as instruments for its work in the world. This is what Gandhi, Mother Teresa, Martin Luther King, and many other spiritual activists understood so deeply, as they implemented spiritual law in their work within the secular and political spheres. This is not to say that one has to become a Gandhi to make a difference, just as one doesn't have to be an Einstein to be a good scientist. Indeed, the transformative principles for social change and cultural evolution that Gandhi and King applied are accessible to us all. By transforming our selves through inner disciplines of consciousness, we become the instruments for a larger wisdom working through us, and this in turn serves the transformation of the world.

There is an amplified power that operates in groups or communities who work with consciousness practices together. Zen master Thich Nhat Hahn has said that the next Buddha will emerge not in the form of an individual,

A key consequence of this worldview is that if we are connected to and led by this larger wisdom of universal consciousness, then our actions and work in the world can become profoundly transformative.

You need only go inward into your own heart, listen deeply there, and live in full integrity with what is revealed to you. This will set you on your true path, and align you with the wisdom of universal love.

but rather in the form of a community of people living in loving kindness and mindful awareness. This is because such a community of people – working together with their hearts and minds in alignment around a shared intention of the highest integrity – creates a powerful field of intentionality that functions like a laser beam of coherent consciousness. Tremendous power can be harnessed in this way from deep within the implicate order which, if carried far enough, can tap into the core of creation itself. Thus rooted in the depths of the implicate order, a group of properly aligned human beings can work directly with the creative process of love itself, and thereby have powerful effects on the unfolding manifestations of consciousness. This is one of many ways that the power of community and the conscious alignment of spiritual groups are becoming increasingly important forces in the new emerging humanity, because they dramatically expand the potentialities and possibilities for human society.

Finally, dear reader, lest you be a bit overwhelmed by all the foregoing talk of worldviews and paradigms, quantum physics and fractal metaphors – rest assured that you don't have to believe this model, or any other theory or philosophy. It has been said, "All models are wrong, but some are useful." Models are just mental maps; ideas to engage the mind – but the real practice is to silence the mind, and go beyond it altogether. You need not agree with or understand anything in this article, nor believe anything in this book, or any other teaching, scripture, or philosophy. You need only go inward into your own heart, listen deeply there, and live in full integrity with what is revealed to you. This will set you on your true path, and align you with the wisdom of universal love.

In closing, the Persian mystic Shebastari tells us, "By love has appeared everything that exists." And further, "By love, that which does not exist, appears as existing." Love is indeed the greatest power in the universe. And divine love is the most powerful form of love. If we give ourselves to the transforming fire of this love, with its attendant demands of radical humility and spiritual surrender, our entire lives begin to burn with passion and longing for the Divine – regardless of the path or tradition we approach it from. This takes us directly into the implicate order, where we reconnect with the Source of all life. At the core of the implicate order is the creative power of love, which initiates us into a mysterious alchemy that opens, from the inside, the inner net of the heart – gateway to the Infinite.

We close with a poem from Rumi, who articulated everything we have been speaking about in this article more elegantly and far more succinctly over 700 years ago:

> *Everything you see has its roots in the Unseen world.*
> *The forms may change, yet the essence remains the same.*
>
> *Every wondrous sight will vanish, every sweet word will fade.*
> *But do not be disheartened,*
> *The Source they come from is eternal –*

Growing, branching out, giving new life and new joy.

Why do you weep? –
That Source is within you,
And this whole world
is springing up from it.
The Source is full,
Its waters are ever-flowing;
Do not grieve, drink your fill!
Don't think it will ever run dry –
This is the endless Ocean!

From the moment you came into this world,
a ladder was placed in front of you that you might escape.
From earth you became plant,
from plant you became animal.
Afterwards you became a human being,
endowed with knowledge, intellect, and faith.
Behold the body, born of dust – how perfect it has become!
Why should you fear its end?
When were you ever made less by dying?

When you pass beyond this human form,
No doubt you will become an angel
And soar through the heavens!
But don't stop there.
Even heavenly bodies grow old.
Pass again from the heavenly realm
and plunge into the vast ocean of Consciousness.

Let the drop of water that is in you become a hundred mighty seas.
But do not think that the drop alone
Becomes the Ocean –
the Ocean, too, becomes the drop!

† Mathematically, the miniature figure shown in Figure 8 is 127 million times smaller than the original shown in Figure 1. And if we kept zooming in on Figure 8, we would find even more miniature replicas. In fact, there are billions of them, embedded throughout the Mandelbrot set, and no two are exactly alike. So the structure of a fractal is something like a set of Russian dolls, where successively smaller dolls are stored inside the larger one, and each doll is painted slightly differently, so no two dolls are exactly alike.

Please see page vi for William Keepin's biography

Evolutionary biologist Elisabet Sahtouris describes how evolution moves from competition to cooperation on all levels, from the smallest bacterial level to human beings to whole societies. A born optimist, she describes how we are in a process of learning cooperation.

Towards a Biomimicry Culture of Cooperation

Elisabet Sahtouris, PhD

Three major crises – in energy, economy and climate – are now confronting us simultaneously, globally, adding up to the greatest challenge in all human history. They are so great, so serious, that nothing short of a fundamental review, revisioning and revising of our entire way of life on planet Earth is required to face this mega-challenge successfully.

This situation, unprecedented in human history, actually makes this an amazing time of opportunity to create the world we all deeply want!

Is that an idle dream, an airy-fairy 'create your own reality' pitch?

Consider: We humans created the reality we have now. It was not imposed on us by fate or any other outside agency. While some may still claim we had nothing to do with global warming, few would deny we have ravaged our planet's ecosystems and loaded our air with pollutants. How many would claim we had no choice in how to produce our energy, or insist that Mother Nature inflicted our money system on us? We humans dreamed up and then realized our economic systems, including our technological path via the exploitation of nature and our focus on consumerism and our extremes of human wealth and poverty. We are an extremely creative species. But something has gone very wrong; something we did not foresee, and we are having very serious trouble understanding and facing that.

If we really look at Nature, we see on the whole that She does not fix what isn't broken. She is profoundly conservative when things are working well, and radically creative when they don't. We would do well to forget our partisan politics and mimic this approach to life's vagaries. Recall that in Arnold Toynbee's classic study of civilizations that failed (1946), the two critical factors proved to be the extreme concentration of wealth and the failure to change when change was called for (Toynbee 1946). These are the current conditions of our global economy in a nutshell, and bigtime Change is now called for.

There were human cultural systems that we created such that they remained sustainable over thousands of years, so why is our most advanced, industrial, hi-tech super-economy, now reaching around the entire globe, proving to be unsustainable in only a few hundred years? To see how this could happen, we must first look at the whole issue of economics.

Economic Basics

What is an economy? I will venture to define the essence of an economy as the relationships involved in the acquisition of raw materials, their transformation into useful products, their distribution and use or consumption, and the disposal and/or recycling of what is not consumed. This definition – and this is very important to understand – is as applicable to our human economy as to nature's ecosystemic economies, as well as to the astonishingly complex economies operating within our own bodies.

Earth has four billion years of experience in economics and may well have something to teach us. Just for starters, nature recycles everything not consumed, which is why it has managed to create endless diversity and resilience, with ever greater complexity, using the same set of finite raw materials for all that time. Furthermore, with us or without us, she is likely to continue doing so for as long as the benevolent sun shines upon her, despite – or perhaps because – she suffers periodic crises that drive her creativity. Let's look at how Earth faces these crises.

As we do, note that Earth's economy is a truly global economy, composed of many and diverse interconnected local ecosystemic economies woven together by global systems of air, water, climate/weather, tectonics, migrations and, not least important, a single gene pool.

Crisis as Opportunity in Nature

We are facing an onrushing Hot Age. Around fifty-five million years ago, Earth had its last Hot Age. In between, since the advent of humanity, our species faced and survived at least a dozen Ice Ages. Only since the last Ice Age have we enjoyed the long – from a human perspective – benign, stable climate in which known human civilizations evolved. It was possible because the last Hot Age plus an Earth-rocking meteor, extinguished the massive reptiles and kicked off a creative wave of mammalian evolution. Crisis for some was opportunity for others in nature's resourceful ways.

In the much older 520 million year-old Cambrian era Burgess Shale, found between two peaks in the Canadian Rockies near Banff, Canada, lies fossil testimony to one of the greatest 'opportunity' responses to crisis in all Earth's history. Interesting that it, too, happened during a time of warm seas and no polar ice – such as we ourselves may be facing – occurring relatively shortly after a 'snowball Earth' climate. In this Cambrian period before land plants and animals appeared, marine invertebrate life reached a fully modern range of basic anatomical variety that more than 500 million years of subsequent evolution has not enlarged. The fossil record of this 'Cambrian Explosion' shows a radiation of animals to fill in vacant niches, left empty as an extinction had cleared out the pre-existing fauna. Once again, crisis for some; opportunity for others.

Let's continue deeper yet into the past. By the Cambrian era, Earthlife had already been through well over half its evolutionary trajectory in years. In fact, for the first half of Earth's biological evolution – for roughly two billion years – archaea (archebacteria) had the whole world to themselves. They evolved amazing lifestyle diversity in their massive proliferation from the depths of the oceans to the highest mountain peaks and even the highest life ever reached in the air, dramatically changing whole landscapes and shallow seafloors as well as the chemical composition of the atmosphere. Their impact is yet to be truly understood outside the halls of science, although they pioneered economic situations and technologies such as harnessing solar energy, building electric motors and developing the first World Wide Web of information exchange we claim as human firsts, as I will describe. (Note our unconscious biomimicry!) My point here is that archebacteria, at the beginning of Earthlife's evolution, were first to make extraordinary responses to global crises – crises of their own making, we should note, unlike the later great extinctions.

The first major such response was to a global food shortage that occurred because the first archebacteria, after spreading all over Earth, were eating up all the free food – the sugars and acids chemically produced via solar UV radiation. Their amazing response was to draw on their own gene pool to change their metabolic pathways such that they could harness solar energy to produce food in the process well known to us as photosynthesis. If we could copy it at a human scale, according to Daniel Nocera at M.I.T., it could fill all our energy needs as long as Earth and we ourselves live. (Note our need for biomimicry in this!)

Before photosynthesis, bacteria had to dwell in seawater or underground, away from burning sunlight. To function in sunlight, the new photosynthesizers were driven to invent enzymes functioning as sunscreens to protect themselves as they lived off the sun's rays and the plentiful minerals and water available to them. Unfortunately, while they did extremely well, they inadvertently created the next big global crisis of atmospheric pollution, leading to the next notable example of taking crisis as opportunity.

Like today's plants that inherited their lifestyle, the photosynthesizing archebacteria gave off oxygen as their waste gas. There were, as yet, no oxygen-needy creatures, so the highly corrosive oxygen, after as much of it as possible was absorbed by seas and rocks and soil reddened by its rusting effects, piled up in the atmosphere in highly significant and dangerous quantities. Along with its direct dangers of killing corrosion, this pollution created the ozone layer which caused further diminution of the old sugar and acid food supply requiring the free passage of UV through the atmosphere.

Once again, life responded with a stunning new lifestyle invention – a whole new way of living using oxygen itself to smash food molecules in the most hi-tech biological lifestyle thus far invented – the one we ourselves inherited from them and call 'breathing'. Bacteria that breathed in oxygen gave off the carbon dioxide needed by the photosynthesizers, thereby completing a give and take exchange in which their plant and animal heirs, including us, still engage.

Life has a dynamic way of oscillating between problems and solutions, which seems to keep evolution happening. The 'breathers' needed food molecules to smash while food was becoming scarcer. Solution: they invented electric motors built into their cell membranes, vastly more efficient than human-designed motors up to the present, attaching flagella to them as propellers. These hi-tech breathers drilled their way into big sluggish fermenting bacteria, which I have called 'bubblers'. (Sahtouris 2000). This initiated the era of bacterial colonialism in which the breathers invaded the bubblers for their 'raw material' molecules. Reproducing by division within the bubblers, they literally occupied them as they exploited and drained away their resources, leaving them weakened or dead. (Is human colonialism biomimicry?)

In this primeval Earth world, we can imagine the many conflicts over scarce food and overcrowding that wreaked havoc, yet simultaneously drove innovation. Eventually, in their encounters with each other, archebacteria somehow discovered the advantages of cooperation over competition: that feeding your enemy is more energy efficient (read: less costly) than killing them off.

Read that last sentence again, because it is the most important discovery any maturing species can make and is very much on our human agenda right now!

All along, in evolving different lifestyles, the archaea had been able to freely trade DNA genes with each other across all the different types in a great World Wide Web of information exchange in which any bacterium had access to the DNA information of any other. Thus they refined a myriad particular cell shapes and lifestyles or roles, such as fixing nitrogen or moving by whiplash propulsion or living in mats of millions.

The crowning glory of all their achievements was the evolution of gigantic collectives with highly sophisticated divisions of labour that became the only other type of cell ever to grace the evolutionary scene: the nucleated cells of which we ourselves are composed. This may have begun, as microbiologist Lynn Margulis and others worked it out, when invading breathers felt their bubbler host weakening and took on some 'bluegreens' (photosynthesizers) to make food for the entire colony. The breathers' motors provided transportation by working in unison on the bubbler's cell membrane to drive the colony into sunlight where the bluegreens could work as needed (Margulis 1998).

In such cooperatives, apparently each specialized bacterium donated the DNA it did not need to fulfil its special function into a common gene library

Life has a dynamic way of oscillating between problems and solutions, which seems to keep evolution happening.

that became the new cell's nucleus. To this day our cells and those of plants, animals and fungi, contain the descendants of these archebacteria in the form of mitochondria (breathers) and chloroplasts (bluegreens).

Nucleated cells went through another billion years repeating the cycle of youthful competition and creativity to mature cooperation in the form of multi-celled creatures. That was the last great leap in evolution – around one billion years ago, bringing us closer to that Cambrian era, when this evolutionary model really took off as described earlier. Ever since, multi-celled creature have been competing when youthful and cooperating when mature.

Maturation Through Crisis

In my view as an evolution biologist, then, the essential pattern in evolution for all species from time immemorial is this very maturation curve from competitive, expansive, youthful economies to cooperative, stable, mature economies. One can see this in what ecologists classify as Type I Pioneer ecosystems and Type III Climax ecosystems today, as well as in looking back over Earth's four billion year history of species' economies.

Some species never make it to maturity. Much of humanity did – but only at the tribal level to which countless human groups matured in cooperation internally and with neighboring tribes, sometimes developing complex economies with large towns and many artefacts, as found at Catal Huyuk in Turkey and many other locations in Africa, Asia, North and South America. Mature cooperation, with other humans as well as with large animals no doubt played a large role in surviving a dozen Ice Ages as humanity did.

In the past 6,000 years or so, we built civilizations – relatively huge socio-economic political systems with complex infrastructures that were mostly internally cooperative despite occasional insurrections. But these mature cooperatives, like the nucleated cell and like the multi-celled creature before them, were new cooperative entities at yet another size scale, and therefore proceeded naturally in the youthful mode of expansionism in competition. Lo, the Age of Empires that shifted over time into national and then corporate empires, had begun!

And so human empires mimic rather well the expansive, competitive phase of juvenile species in nature from the original archaea (bacteria) to the grasses that evolved along with humans and are also still in that juvenile take-over, make-over whatever you can to stay in the game mode Darwin described so well. Interesting that humans and those youthful grasses – in the version humans call 'grain' or 'corn' – have come to depend on each other.

Yes, Darwinian evolution describes the juvenile phase, and that is precisely why the entrepreneurs of our Industrial Age loved that theory as much as the Soviet Union loved Kropotkin's version of evolution, titled Mutual Aid, all about the cooperative phases of species evolution, which rationalized collectivism. In the first, community was sacrificed to the individual's interest; in the latter the individual's interest was sacrificed to that of the collective. Two half theories that make a whole when put together and make the connections between the ecologists' different types of ecosystems. The learning curve of maturation ties it all together in an elegant whole.

The recognition that our current way of life is unsustainable (literally implying we must live differently) is a new and vital insight, without which we could not see any need to change the way we live on what seemed like a limitlessly provident planet, now so obviously ravaged by our youthful empire building to a critical point, if not already beyond it.

All our technology has come through biomimicry – from spinning like silkworms and weaving like spiders, building like termites and tunneling like moles, flying like birds and computing like brains, to using radar like bats and sonar like dolphins, and so on and on. But now it is time for the biggest and evolutionarily greatest biomimicry feat of all: copying those of our ancestors who made it to mature sustainability, pulling back on our economic expansion just as our bodies did when reaching mature size and shifting to maintaining stable sustainability.

Looking at our recent history, we see many experiments in cooperation pushing us to our truly global cooperative maturity: from the United States of America to the European Union, from NATO and SEATO and other alliances to World Parliaments of Religion, a World Court and International Space Stations, from VISA cards crossing cultures and currencies to International Air Traffic Control, and so on and on.

The Internet is the largest self-organizing living system created by humanity and is changing everything. The top-down hierarchies that worked to maintain and expand empires are giving way to democratic and even more mature living systems ways of organizing and governing ourselves; even the gifting economies arising all over it, as well as in local communities, biomimic mature species economics.

If there is one biological system that can give us the clues in an up close and personal model available to us all, it is our own bodies. There is no more amazing or mature economy to mimic as we design our own future than the bodies in which each and every one of us, regardless of political persuasion, is walking around – bodies in which no organ either exploits the rest for its own benefit or interferes with diversity by trying to make the others more like itself.

Each of your up to one hundred trillion cells has some thirty thousand recycling centres in it just to keep all those proteins you are made of healthy. Each of those is as sophisticated as a chipper machine would be if you could stick a dead or damaged tree into one and get a healthy live tree out the other end instead of a pike of chips! And they exist along with a thousand mitochondrial banks in each cell, giving out free ATP stored-value debit cards 24/7 with no interest, not even pay-back of what you spent – a currency system we could well biomimic as soon as possible in place of our

All our technology has come through biomimicry – from spinning like silkworms and weaving like spiders, building like termites and tunneling like moles...

wealth-concentrating debt money.

It has become clear to me that the mature cooperative phase of species is often driven into existence by crises and I am happy to note how the vast majority of humans becomes highly cooperative in times of disaster, surviving predations of the very few to create wellbeing for the many. It is in our genes, our blood and bones, to cooperate. We have been through this before, just never before at a global size level.

Species that become sustainable – that survive a really long time – get to their mature collaborative phase while others, stuck in adolescent behaviours that no longer serve them, die out. Humanity now stands on the brink of maturity in the midst of disasters of our own making. Let us take heart from our most ancient Earth ancestors, the archea – the only other creatures of the living Earth to create global disasters through their own behaviour and solve them. Let us see if we can do as well as they did! Let a mature and cooperative global economy be our goal and let us make it as successful, as efficient and resilient, as our own highly evolved bodies.

The global economy we built as a resource-rapacious, competitive monopoly game based on debt money and powered by fossil fuels was a necessary youthful phase. We are ready now to leap into maturity. We the people can declare our solidarity with each other around the globe, stop making war on each other, roll up our sleeves, and do the positive work needed to develop clean energy sources, move coastal cities uphill, reinvent money, green deserts, and cooperate in all our cultural and religious diversity to build a world that works for all, whether or not our governments follow our lead.

As Rumi asked: Why do you stay in prison when the door is so wide open?

Elisabet Sahtouris PhD (www.sahtouris.com) is an internationally known evolution biologist, futurist, author and speaker living in Mallorca. With a post-doctoral degree at the American Museum of Natural History, she taught at MIT and the University of Massachusetts, contributed to the NOVA-Horizon TV series, is a fellow of the World Business Academy, and a member of the World Wisdom Council. Her venues include the World Bank, UN, Boeing, Siemens, Hewlett-Packard, South African Rand Bank, Caux Round Table, Tokyo International Forum, the governments of Australia, New Zealand and the Netherlands, Sao Paulo business schools and State of the World Forums. Author of EarthDance: Living Systems in Evolution; A Walk Through Time: From Stardust to Us; and Biology Revisioned with Willis Harman. Her website is **www.sahtouris.com**, where several books and many articles can be downloaded for free.

Ross Jackson explores the factors that stultify our capacity to evolve towards an ecological or Gaian global civilisation. He presents a practical vision of how we can change our global political and economical systems to transform the world.

A Gaian Worldview

Ross Jackson

Imagine how a global society might look – politically and economically – if we are able to evolve into a truly sustainable and equitable global civilization based on the emerging holistic worldview, which I call the Gaian Paradigm. How would it differ from current society and why? What are the major barriers to be overcome if we are to make any progress towards the desired utopia? What are the deeper causes that are hindering movement in the right direction? In the following, I will summarize my answers to the above questions and conclude with the broad outline of a proposal for a possible political initiative, which I call the breakaway strategy, which hopefully will be able to surmount the identified barriers and thus put our civilization on a trajectory that can lead toward the desired long term goal.

The Global Crisis

Our planetary civilization is in a very precarious and vulnerable state at this time as several global threats confront us, some of which could be fatal for humanity if not dealt with. Global warming is probably the threat, which most people immediately think of in this regard. The reality of the threat to our climate, caused primarily by burning fossil fuels, has not only broad scientific support, but the message has been received and accepted as real by a growing majority of world citizens and political leaders. But other threats are just as important and not yet widely recognized.

Even if we were able to magically resolve the global warming problem immediately with the wave of a magic wand, we would still be faced with the just as critical issue of over-consumption of our natural capital (water aquifers, topsoil, micro-organisms, and biomass). Our so-called 'ecological footprint' measures the number of hectares of land that would be required by the population of a region in order to provide the renewable resources consumed and the sinks to absorb waste products. Globally, roughly 2.7 hectares per person was required in 2005 according to the most recent measures from WWF.[1] However, the available land and sea space was only 2.1 hectares per capita for a population of 6.6 billion. This means we had an over-consumption or 'overshoot' of about 30% in 2005, and the overshoot is getting higher each year as population and economic growth increase. The distribution of this overshoot is skewed with a heavy overweight to the most industrialized countries. This situation is clearly not sustainable. We are consuming the natural capital that sustains us. No one knows for sure how much overload the ecosystem can tolerate before it collapses, but a continuation of current policies will lead to certain disaster sooner or later. Nevertheless, rather than deal with this threat by reducing consumption, every country is doing precisely the opposite by adopting more economic growth as its major political goal.

At an international conference on fertility in June, 2007, one of the world's leading fertility researchers, Niels Skakkebæk of Denmark, sounded an alarm concerning the phenomenon of declining male fertility due to estrogen-like endocrine disruptors and pesticide residues in mother's milk, a threat he declared was every bit as serious as global warming.[2]

A leading British expert in biochemical genetics, Mae-Wan Ho, has published a similar dire warning on genetic engineering technology, as scientists have removed the barriers to virus recombination and gene transfer that have been closed for millennia, an opening, which could in the worst case, she claims, threaten human existence. She calls the field an 'unprecedented alliance between bad science and big business'.[3]

While the above four threats are all potentially fatal for humanity, a different kind of threat, while not fatal, could prove far more important in the short run, namely Peak Oil. Oil geologists are in general agreement that global oil production will peak soon and go into permanent decline, while oil demand is inexorably increasing due to the political focus on economic growth. Demand will soon exceed supply for the first time ever. The danger is that exploding oil prices will throw the global economy into a long period of chaos, recession and negative growth.

How Did We Get Into This Mess?

I claim that all of the above threats can be traced to the same underlying cause. We are experiencing the shadow side of a centuries-old worldview that separates Man from Nature. This worldview sees the world as a mechanical machine made up of individual parts that can be manipulated separately. Nature is something outside of us, having no intrinsic value, and is to be conquered. This reductionist way of looking at the world has been the foundation of modern science and has provided us with an enormous increase in living standards. It is generally considered a great success by most people. However, we are beginning to realize that there were many hidden costs, and the bills are now coming due.

As long as we were relatively few people and our technologies were relatively harmless, the future seemed to be an unending horizon of growth and material progress. But in recent years we have begun to meet the limits to growth on a finite planet while simultaneously developing powerful technologies with far-reaching effects. Now we can mine minerals by blowing the tops off mountains and fish by violent scraping of the seabed, destroying much sea and plant life indiscriminately.

We are collectively learning an important lesson in these years, namely that the ecological system is far more complex and unpredictable than we ever realized. A seemingly rational development in one area, burning fossil fuels, results in a threat elsewhere, in climate change. A shift from natural to synthetic materials, while apparently cost-effective, results in a threat to fertility from something as simple as the plastic lining in a can of beans. Genetically engineered plants may be more robust, but at the high cost of disabling nature's anti-virus defenses. We are discovering that we are an integral part of a living organism – Gaia – here every component is interconnected in ways we do not fully understand.

And yet, our political system seems unable to take coordinated global action to deal with the multi-dimensional crisis facing us. Why? Because our political organization reflects the same separatist worldview. Every country looks out for its own interests. There is no global governance, no international institution that has the interest of the whole planet as its mandate and the power to do something about it.

...our political organization reflects the same separatist worldview... There is no global governance, no international institution that has the interest of the whole planet as its mandate and the power to do something about it.

Neo-liberal Economics

A worldview of separation leads to one-dimension thinking – a focus on one aspect at the exclusion of all else – in reality a form of extremism. Modern economics is an example of this, particularly neo-liberal economics, which has dominated the field since the 1980s. In neo-liberal economics there is no room for ecological or social considerations. Nature is seen as a free resource to be exploited without restriction. Social effects are 'externalized' off the corporate balance sheet. Capital is seen as something which should be free to move anywhere at a moment's notice without restriction. Shareholders of private corporations are seen as the ultimate rulers, free to do as they like to maximize their profits in a world of 'free trade', privatization and deregulation, the basic claim being that this will benefit everyone. In practice, as shown by independent studies, it benefits only the already wealthy, in both the industrial and developing countries, at the expense of everyone else and the environment. In my opinion, if our goal was to drive our global civilization to ecological ruin, we could not find a more efficient way to do it than by inventing neo-liberal economics.

The Growth Fallacy

In spite of documented global over-consumption, every country has the political goal of maximizing growth. This is a hopeless mission. If all countries were to achieve the same level of consumption as the USA, the overshoot would be 360%, corresponding to the need for almost four additional planets. Clearly Humankind would be dead along with the ecosystem long before we got to that point. As any biologist will confirm, growth always ends with either collapse or a steady-state 'climax' state, such as a rain forest. As a civilization grows, the costs of more growth will always increase, since we implement the least costly solutions first, and the more expensive ones later. At some point the marginal costs will exceed the marginal benefits. It is at this point that many past civilizations began to collapse according to historian Joseph Tainter.[4]

A number of alternative economists have begun measuring the real benefits to society of economic growth by deducting from Gross Domestic Product (GDP) the components which make no positive contribution to our well-being but are more like negative byproducts (pollution cleanup, carbon sequestration, highway accidents, exploding health costs, etc.) We see the result below for the USA in the period 1950-2002 for one such measure of the net benefits, called the Genuine Progress Indicator.[5]

Note that the net benefits went into decline around the 1970s. Since then, economic growth has had no net benefits to American society. The authors estimate that the American GDP in 2002 overestimated the real benefits by $25,000 per person. A similar pattern has been measured in several other countries, including the UK, the Netherlands, Germany, Austria and Sweden. Unfortunately, we may already be in the midst of a collapse of our civilization.

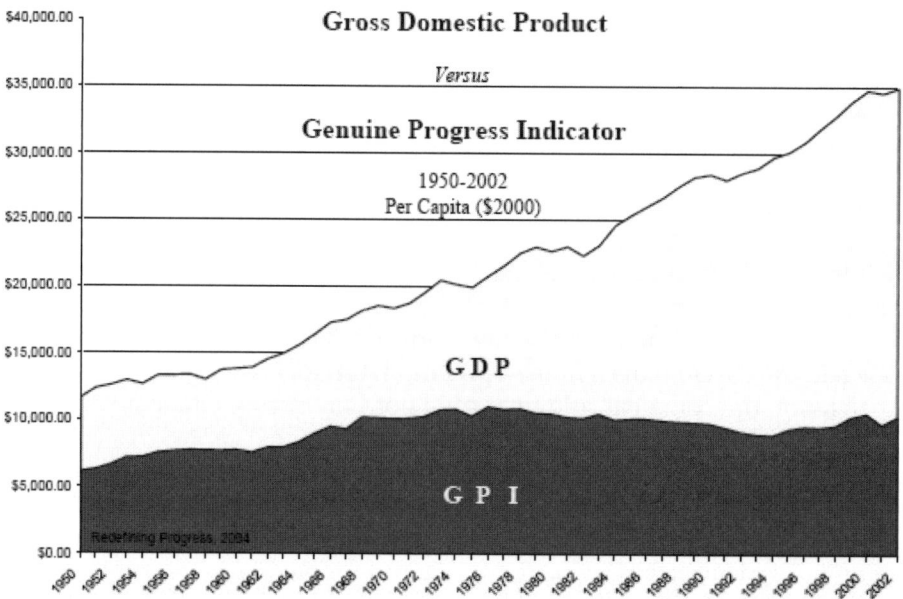

The Genuine Progress Indicator.

The WTO and Global Governance

The prime institution that implements neo-liberal economics is the World Trade Organization (WTO), which replaced the former trade regime, the General Agreement on Tariffs and Trade (GATT) in 1995 without any public debate and despite the protests of the developing countries. The major differences were the introduction of binding rules to resolve conflicts between nations and a shift in control from nation states to foreign commercial interests. It has been rightly called a corporate charter, written by corporations for corporations. In the WTO there is no place for ecological or social voices and no democracy, with the developing countries cast in the role of providers of cheap labor and cheap raw materials for the rich industrialized countries, who reserve for themselves the right to give enormous subsidies and protection to their domestic industries. The WTO is the closest thing we have today to an institution of global governance.

The Gaian Paradigm

If we are to go to the root of the problem, and look for solutions, we must begin by discarding the old mechanical worldview – sometimes called the Newtonian/Cartesian paradigm after Isaac Newton and René Descartes. We must replace it with one that better reflects the realities of planetary life. Fortunately this is happening gradually. A new, holistic worldview of inter-connectedness and solidarity is emerging. We can call it the Gaian Paradigm for reference, reflecting the recognition of the earth as a living organism with humanity as an integral part, and having a very special role. Such a worldview must be all inclusive. We can no more reject a minority of human society, or animal or plant life than we can reject a part of our human bodies.

The death blow to the old worldview can be traced to the 1920s when quantum physics theory indicated that all particles throughout the universe were 'entangled' or interconnected, thus falsifying Descartes' idea that we can separate the observer from the observed. Thus we cannot separate Man from Nature, as René Descartes believed. Everything is inter-connected in unpredictable ways. It takes many decades for such an insight to work its way through society, but it is happening slowly. In every field of endeavor, a small minority of forward thinking people have made the shift. How might things look when society as a whole has made this shift?

A Gaian Society – One Hundred Years Hence

A global society based on a Gaian worldview would be very different from today's world. Let us now imagine how things might be in a world that was organized according to the principles of the Gaian Paradigm, say one hundred years from now.

Gaian Economics

Economics is very central to the various crises facing the world in the early 21st century. When we do not take account of ecological and social costs, and when we allow enormous subsidies, then we distort the whole pricing structure and take completely wrong and potentially disastrous decisions. Furthermore, any Life-based view of society recognizes that there are many aspects of life that are beyond the market economy, even with correct ecological pricing, for example, the informal sector and social needs. So a market society is not an acceptable solution. On the other hand, rational beings will respect that the free play of self-interest often leads to far more innovative and more efficient solutions than any centrally planned scheme can achieve.

So in Gaian society, economics will be neither capitalist nor socialist, but 'Gaian'. No raw capitalism will be found here, and no inefficient central planning either. Corporations will be allowed substantial freedom to find optimal solutions, but within the framework of a rule set that gives incentives for innovation that is protective of the environment and provides penalties for the opposite, for example a tax on CO_2 emissions. A similar framework for social needs will be required – for example, a global minimum guarantee on basic necessities of every human, including clean water, soil and air and membership of a social network. Demands will be put on corporations by law, for example to live up to their social responsibilities and to respect all stakeholder interests.

... in Gaian society, economics will be neither capitalist nor socialist, but 'Gaian'. No raw capitalism will be found here, and no inefficient central planning either.

Localization

Gaian society will be characterized by a shift from centralization to localization, for a number of reasons. One driving factor will be the post Peak Oil energy descent. The high price of oil is going to have a major impact on industrial agriculture and transportation, both of which are very dependent on cheap oil. Escalating food prices and transportation costs as well as health considerations will result in an explosion of local food production all over the world and far less travel. Increased local food production will bring with it a number of secondary industries and create a revival of local communities as the backbone of every nation. This will be further supported by the shift to decentralized energy sources, in particular solar energy, and a migration from the cities to the countryside, as citizens seek out a better life in ecovillages and other well-functioning local communities. Local democracy will blossom as power shifts from centralized institutions to self-reliant communities and regions.

Let us now consider briefly some of the major political/economic institutions that might evolve in a Gaian society.

The Gaian League

My vision of the ideal Gaian world would include as its centerpiece an international organization in which all sovereign states were members. Let us call it the Gaian League, an organization designed to protect and further the interests of all world citizens, not as a centralized world government, but as an organization coordinating certain activities of hundreds of sovereign states, who otherwise run their countries as they see fit without outside interference in a world of diverse cultures. The areas where member states are required to cede a limited degree of sovereignty are two only – ecological sustainability and respect for human rights. Only in this way can the indefinite continuance of the species be guaranteed while respecting social justice. To carry out its goals, the Gaian League will have to form new institutions, including, but not limited to, the following.

The Gaian Congress (GC)

A new international legislature, with a major focus on ecological sustainability, will be established. Let us call it the Gaian Congress, in which each sovereign state is represented, and binding resolutions can be taken by a qualified majority. The resolutions and directives adopted would, by definition, constitute the international law to be applied within member states. This could become the first attempt at a formal set of international laws with economic consequences.

This could be made possible by the introduction of an 'Economic Sanctions Protocol', for example, that would override not only national legislation, but any agreement reached by any international treaty, including trade agreements. In practice, the Economic Sanctions Protocol would mean that a nation could demand and get compensation for any loss due to non-compliance of another state. Of course, the magnitude and duration of the sanctions and compensation must be reasonable and appropriate.

The Gaian Trade Organization (GTO)

In the WTO regime, sovereign states have in effect given away a portion of their sovereignty to the WTO, whose trade experts can interpret and enforce the rules in any conflict. These binding rules include no demands on corporations, allow foreign companies to sell their products without disclosing how they were produced, place the burden of proof of health risks on the consumer, and give foreign companies the right to any advantage given to a domestic company. It should be clear that these rules have to be reversed if sovereign states are to have any real control over their countries, not least as regards their economies, food security, the environment and the social costs of corporate decisions.

In Gaian society, trade will be organized in keeping with the Gaian Paradigm. For the sake of reference, let us call the new organization the Gaian Trade Organization (GTO). This entity will be based on a set of principles consistent with the needs of a truly sustainable civilization. The major principles of the GTO can be stated very simply, and are basically a reversal of three WTO/IMF principles back to what was the norm during the extremely successful GATT regime prior to 1995:

- Sovereign states have the final word in trade matters.
- Capital controls are reinstated on investment flows.
- Developing countries are once again positively discriminated.

Corporations will function freely within the rule set laid down by the GTO that protects the environment and social structures. Rationing of resources – for example putting a ceiling on permissible global CO_2 emissions – and more correct ecological pricing will stimulate the development of far more efficient production technologies. New technologies will emerge based on the principle of learning from nature how to produce with little energy and at room temperature and with no waste. Close to 100% recycling will become the norm in order to not take more from the earth than is replaced.

In the GTO regime, there will be no general binding rules regarding economic activity, only voluntary guidelines. Of course, bilateral and multilateral agreements can and will be entered by GTO members with other members and non-members, and these may involve binding rules and binding arbitration on an individual basis, but nothing can be forced upon a sovereign state.

The Gaian Development Bank

This institution, owned by the members and financed by the pooling of foreign exchange reserves, would replace the World Bank and the IMF. Its primary activity would be to establish a network of non-profit banks in the developing countries with the goal of making local currency loans and investments to promote the real development of self-reliant sustainable nation states that can produce their basic necessities themselves with a minimum of foreign interference.

The Gaian Council

A Gaian Council consisting of a few highly respected individuals elected from major earth regions will fulfil the role of a global governance institution having the mandate to act in the interests of the entire global community. Eligibility for election will be determined by demonstrated dedication to the higher goals of humanity and a level of high spiritual evolution. Gaian Council's formal power will rest in its right to veto any resolution adopted by the Gaian Congress if not seen to be in the common interest. The Gaian Council's mandate will also be to protect all minorities, however small, and to mediate conflicts between regions or nation states.

The Gaian Council will be neither an executive, nor a legislature, nor a court. Its role is to resolve conflicts among the member states and to give expression to the desired principles and general direction of the development of our global civilization, based on widespread consultation. In particular, its task would be to ensure a sustainable future while respecting human rights. In practice, its formal powers would ideally be used sparingly. The combination of its veto power and guiding principles, together with enforceable directives on sustainability practices and human rights by the Gaian Congress should provide the necessary tools to guide global society.

The Gaian Council should not be confused with a world government. A centralized world government would be something quite different and would most likely be disastrous, as it would only lead to hegemony by a particular group strong enough to further its own self-interest. Gaian society will be far more decentralized than is imaginable today with many small cooperating nation states having diverse structures, cultures and traditions akin to nature's own organization. However, centralized protection of the environment, universal human rights and the rights of minorities is a different matter, and would define the Gaian Council's most important functions.

How Do We Get From Here To There?

It is so unlikely as to be virtually impossible that any existing organization or group of powerful states, such as the G20, the OECD or the United Nations could take the lead in promoting a world order similar to the Gaian society described above. Besides the enormous difficulty in reaching any kind of agreement among so many diverse players, such an initiative would conflict with the perceived self-interest of the most powerful states. They are locked into defending the status quo, even though it may lead to the worst possible result and disaster for all, including themselves—a situation known in game theory as the 'prisoners' dilemma'.

But there is another possibility. We know that throughout the world, in every country and in every walk of life, there are people who would support a visionary political initiative based on the Gaian Paradigm. Many are found in the thousands of NGOs that promote various single issue themes. Many are found in life-style movements, like Voluntary Simplicity, Engaged Spirituality and the Ecovillage Movement. But many are also found in academic circles and in the business world, and among ordinary citizens in the streets, such as the Occupy movement. They are everywhere; not least in the most powerful states, such as the USA and the EU. They are a minority, but a large minority, and a growing one. Sociologist Paul Ray estimates that about 35% of the populace in Western states, the so-called 'cultural creatives', share values that are very similar to those of Gaian society.[6] Could these minorities be mobilized? Perhaps.

The Breakaway Strategy

My recommended strategy is in principle very simple. A small group of nations initiates a joint political initiative. They leave the WTO and announce that they are forming a new organization, the Gaian League, based on the principles of the Gaian Paradigm. They then form new institutions as part of the Gaian League, including the Gaian Congress, the Gaian Trade Organization, the Gaian Development Bank and the Gaian Council, as described above, introducing for the first time a formal system of international law with meaningful consequences for member states. The initiators must emphasize that the Gaian League is intended to meet the needs of a global civilization based on democracy, sustainability and justice, and not just their own self-interests. Other states will be invited to join when they feel they are ready.

The breakaway initiative will require considerable dialogue and planning before implementation. There are many aspects to consider, including the risks of negative reactions from abroad. No country is going to move beyond the dialogue stage without careful consideration of all aspects of the decision. So the very first step will be an informal dialogue among a small group of nations who are willing to take a closer look at the concept.

It is pure speculation to guess from where a positive response to the breakaway proposal might come. A lot will depend on personalities. A very visionary and charismatic leader of high integrity could make all the difference. Some of the smaller countries of northern Europe have the ideal combination of a high level of global consciousness and the ability to afford it. But are they courageous enough to take the step? A country like Denmark, my adopted home, ought to be in this group. However, it is unlikely at his time due to the EU. Denmark, as well as Sweden, Finland and the Netherlands, would have to leave not only the WTO, but also the EU in order to join the GTO, as EU membership prevents a member country from using tariffs to protect the environment and other national interests. That is probably too big a step to take at this stage for these countries. However, Norway and Iceland are interesting possibilities. Iceland has the declared goal to be the first country to be completely independent of oil (thanks to geothermal energy sources) and is working consciously to develop a more 'green' profile.

In other parts of the world, countries like Costa Rica (no army) and Bhutan (maximizing Gross National Happiness) have shown that they share many Gaian Paradigm values. In Africa, Senegal's government was the first anywhere to actively support the ecovillage movement. In Sri Lanka, the Buddhist Sarvodaya movement links 15,000 villages in a movement sharing many Gaian Paradigm values. In South America, thirty three states recently formed a regional alliance, CELAC, which has objectives similar in many respects to the GTO.

If the initiative is launched properly, it will receive the support of the cultural creatives of the world. Indeed, grass roots support is critical to success. Demonstrators in the streets of non-member states must insist on a referendum on the question of joining the Gaian League, even in opposition to their elected leaders. In time, other nations will hopefully join as the flaws of the old paradigm become more and more obvious. They will then contribute to the vision and help to move the world in the direction of a truly sustainable and just global society.

References

1. 'Living Planet Report 2006'; World Wildlife Fund; see www.panda.org
2. 'Forsker advarer: Som art er vi I fare'; Politiken; Copenhagen, 1 June, 2007.
3. Mae-Wan Ho; *Genetic Engineering, Dream or Nightmare*, 2nd edition; Gateway, Dublin, 1999.
4. Joseph Tainter; *The Collapse of Complex Societies*; Cambridge University Press, 1988.
5. 'The Genuine Progress Indicator, 1950-2002 (2004 update)'; www.redefiningprogress.org/publications
6. Paul H. Ray and Sherry Ruth Anderson; *The Cultural Creatives*; Harmony Books, New York, 2000.

Ross Jackson, PhD, was born in Canada, but has lived most of his adult life in Denmark. His formal education includes degrees in Engineering Physics, Business Administration and Economics. He was for many years an IT consultant with his own firm, later specializing in international finance and currency markets. He is founder-chairman of Gaia Trust, established 1987, a Danish-based charitable association that has been a prime financial supporter of the Global Ecovillage Network and Gaia Education as well as hundreds of other ecological sustainability projects worldwide. His books include *And We ARE Doing It: Building an Ecovillage Future; Kali Yuga Odyssey: A Spiritual Journey; Shaker of the Speare: the Francis Bacon Story*; and his most recent, *Occupy World Street: A Global Roadmap for Radical Economic and Political Reform*. His website is **www.ross-jackson.com**

May East explains why an important aspect of evolution is the restoration of the feminine principle of creation in complementary relation with the masculine, manifesting a new mythology of the universe.

The Dual Origin of the Universe

May East

Incredible, mysterious, inspiring. As many stars as the Milky Way as many ways to describe the limitless Universe in which our world abide, the matrix of the Great Mystery.

The self originating Universe, continuously self creating, non-local and indivisibly whole. The universe, the sum total of all states of consciousness and activity, hidden by the barriers of space and time, it has existed for ever beyond our reach, unknown, unexplored.

Now at long last, those cosmic barriers have began to lift, and we have the first glimpses of those once secret domains. And what we have seen has left us astounded... the first faint comprehension of the universe so bizarre and bewildering is challenging the very notions of matter and energy. It is challenging the very core of our worldviews, belief systems and also our rationality.

The finest cosmologists of our times believe the universe began with a great bang, about 10 to 20 billions years ago. When a primordial mass exploded in a titanic holocaust. This fiery ball gradually cooled as it expanded outwards. Giant clouds of swirling gases formed the celestial bodies. Entire galaxies took shape. This world of the big bang, also known as the big birth, may expand forever. And from within this out-breathing of Life force, two major streams of living energy manifested, informing, since then, all Life, seen and unseen.

The two cosmic currents referred by spiritual teachers, seers and intuitives of all times as the Dual Origin of the Universe. The feminine principle of creation, the masculine principle of creation. The two cosmic currents, known as yin and yang by the Chinese and identified as wave and particle, fermeon and bozon by our contemporary physicists, permeate the very fabric of our Universe.

Our generation is gradually realizing the majestic cosmic law of equivalency, the law of the Dual Origin, as the foundation of existence. The predominance of one origin over the other has created a lack of balance and destruction, which may now be well observed in all of life. Wise ones such as Vivekananda affirmed, "The bird of spirit can only fly with two wings," and Helena Roerich said, "How can priority be given to one energy over another when the fiery tension can occur only in fusion? The acknowledgement of the two Origins is the foundation of Cosmos."

From times immemorial the existence and myth of the Mother of the World informed the history of humanity. Myth, the secret door, the mysterious opening, through which the energies of the cosmos pour into human cultural expression. "Her hand traces an unbreakable thread."Throughout the civilisations the feminine principle of creation has taken many faces.

The ancient ones would say: The Mountain of the Mother extends from the Earth to the Heavens, indicating thus the unity of all that exists. It seems that humanity's first images of divine power were great mother goddesses with wide hips, fertile bellies and full breasts, concentrated in the drama of birth, nourishment and fertility. The story of a great primeval goddess is told in the ancient Palaeolithic caves, the most sacred of place, the sanctuary, the womb and the source of The Great Mother's regenerative power.

In Europe the first images appeared across a vast expanse of land stretching from the Pyrenees to Lake Baikal in Siberia, suggesting there was a continuity of a religious structure from what is known today as France to Russia. Recent discoveries in the Amazon basin showed that the whole region was home to an affluent and sophisticated society that worshipped the divine feminine where women were leaders, and the interconnected world was experienced as a constant flux.

The Mother Goddess in the Neolithic is in an image that more obviously than before inspires a perception of the universe as organic, alive and a sacred whole, in which humanity, the Earth and all creatures on Earth participates as Her children. As the Great Mother, she presides over the whole creation as goddess of life, death and regeneration, containing within herself the life of plants as well as the life of animals and human beings. The Divine Mother in all her manifestation was the symbol of the unity of all life in Nature. Her power was in water and stone, in tomb and cave, in animals and birds, snakes and fish, hill, trees, flowers and grains.

In the Bronze Age, the Divine Mother was the basis of the cults of Inanna of Sumeria, Cybele of Anatolia, Astarte of Cannan, Athena of Crete. The divine Feminine was manifested through the pantheon of Greek Goddesses and of Indian Goddesses: Aditi, Parvati, Durga Saraswati, Kali. She was revered as Tara, Tian Hou, Fuji, Hu Tu, Yu Nu, Kwan Yin, in the East. In the Brazilian Guarani tradition, she was know as Aracy, the mother of the day. The highest manifestation of the Feminine Principle has been called by many names, among them the Mother of the World, Mother of the Universe, Ishtar and Sophia. She was also known as the Woman clothed with the Sun, Amaterasu. Isis was the greatest Goddess in Egypt and was worshipped

Ethnic tribal native prehistoric priest symbol.
© *Artemiy Bogdanoff/Shutterstock*

for over 3000 years until the second century AD when her cult and many of her images passed directly on to the figure of Mary. Mary is the unrecognised Mother Goddess of the Christian tradition. To the Gnostic Christians she was known as Holy Spirit, one of the Divine Triad, while ecclesiastical Christianity has regarded the Holy Trinity as entirely masculine.

The web of space and time that the mother goddess once spun from her eternal womb – from Neolithic Goddess figures buried with spindle whorls, through the Greek spinners of destiny, down to Mary – had become the cosmic web in which all life was related. The banner of peace held by Madonna Oriflama suggests the meeting of the past, the present and the future within the ring of eternity held by the divine feminine.

Many of the mother goddesses were born from the sea, from the Sumerian Nammu, the Egyptian Isis, the Greek Aphrodite down to Mary (whose name in Latin means sea). Now this image had come back into the imagination as 'the ocean of energy' of the implicate order of the New Physics. (See page 7)

So where today we can find the myth of the Eternal Feminine? If we turn to the discoveries of the New Sciences it appears astonishingly, as if the old goddesses are re-emerging in a new form, not as a personalised image of a female deity, but as what that image represents a vision of life as a interconnected whole in which all life participate in mutual relationship and where all participants are dynamically alive.

Discovering the Feminine Principle Within

The tale that has most deeply informed my life was the one of my great great grandmother who was a free Guarani woman, living in deep intimacy with the subtropical forests of Brazil until the day she was lassoed by my great-great-grand father, an European from the Iberia peninsula. She was uprooted and given a Christian name, Maria.

How uncomfortable it was to hear the elders of my family sharing this story in between laughs and jokes. As I grew up I gradually learnt I could identify and tap from both lineages: the oppressor and the oppressed, the adventurer and the captive. Yet the evolutionary pull within invited me to go beyond the polarised roles and use my mixed blood, as a gift for bridging the worlds, igniting my search for the Feminine Principle of Creation.

In this quest I went from the beginnings of time, journey through the eras, finding some women as forerunners of their times. They are not considered forerunners because of personal achievements, but for the effect their efforts have had on the lives of countless others. From daring feats of bravery to the ways of a compassionate heart, from clear thinking to unexpected acts of beauty and inclusion, they have advanced womanhood in the world, they have been catalysts of change. They are what I name, pearls in the necklace of the Mother of the World.

Helena Blavatsky: "Knowledge increases in proportion to its use; that is, the more we teach the more we learn."

Eleanor Roosevelt: "It is not fair to ask of others what you are not willing to do yourself."

Alice Bailey: "If you are inspired, you are bound to inspire others, otherwise you blow up and bust."

Susan Anthony: "The day will come when men will recognise woman as his peer, not only at the fireside, but in councils of the nation. Then, and not until then, will there be the perfect comradeship, the ideal union between the sexes that shall result in the highest development of the race."

Mother Teresa: "We can do no great things, only small things with great love."

Marie Curie: "You cannot hope to build a better world without improving the individuals. To that end each of us must work for his/her own improvement, and at the same time share a general responsibility for all humanity, our duty being to aid those to whom we think we can be most useful."

Margaret Mead: "We are living beyond our means. As people we have developed a life style that is draining the earth of its priceless and irreplaceable resources without regard for the future of our children and people around the world."

Inspired by these women and many more, I encountered ecofeminism in the 90s, a social movement that emerged out of women's liberation, and the political and environmental movements which appeared simultaneously in the 1960s, and continued to grow in influence throughout the 70s and beyond. The word ecofeminism was first used in 1974 by French writer Francoise D'Eaubonne, in her book *Ecofeminism or Death* talking about women's potential to bring about an ecological revolution. The fundamental thesis of ecofeminism was that the domination and destruction of nature and the oppression of women come from the same root cause, the same worldview.

Ecofeminism calls for an end to all oppressions, arguing that no attempt to liberate women, or any other oppressed group, will be successful without an equal attempt to liberate nature. Ecofeminist action has emerged in the South over the same period as in the North. But it is only recently that women in the North and South have come together, and the global women's and environmental movements have become a real force for change at local, regional and international levels.

An example is the Amazonean Movimento Fraterno das Mulheres Lutadoras de Anapu, the Fraternal Movement of Mobilised Women of Anapu. This has been working since 1995 to help reverse some of the harmful effects of development in Anapu, a municipality of the Amazon,

"Knowledge increases in proportion to its use; that is, the more we teach the more we learn."
Helena Blavatsky

focusing primarily on environmental sustainability approaches and gender equity issues.

Women from Anapu have fundraised internationally and, with the grants received, they are recovering degraded agricultural areas by re-introducing native plant species and recovering the biodiversity. They are involving both women and men in their activities to conserve their land, which in turn will help prevent further migration and forest degradation. The work has so far resulted in greatly increased knowledge and understanding among producers that the degradation of forested areas leads to further impoverishment. They are fierce guardians of the genetic diversity of the forest.

This moment in history is calling forth the best and the strongest in each one of us. The great turning away from unsustainable industrial growth society to sustainable communities, and the rebellion against women's intellectual, economic and social marginality in the South, are occurring with a force of spring floods, breaking out of rock and ground in different places and in a great variety of courses.

With the restoration of the feminine to a complementary relation with the masculine, a new mythology of the universe as one harmonious living whole is emerging. Thus the Cosmos exists in the greatness of the Dual Origin. We are restoring the sense of wonder and awe as we expand and deepen the knowledge of the Universe. The universe again become sacred and luminous. The universe and its wave particle structure. The universe the high abode of the feminine and masculine principles of creation.

Extracted from May East's Winter School of Ecofeminism inspirational lectures.

May East is a Brazilian social change activist who has spent the last 30 years working internationally with music, indigenous people, women, anti-nuclear, environmental, ecovillage and transition movements. She is the programme director of Gaia Education, leading a whole generation of sustainability designers and educators delivering trainings in 30 countries in the most different stages of development and in both urban and rural contexts.

Hildur Jackson describes the values of the new worldview and the global consequences they could initiate. These values are intrinsic to the ecovillage movement.

Living the New Worldview: Global Justice & Saving Three Billion Years of Evolution

Hildur Jackson

At the UN Habitat conference in Istanbul in 1996 I interviewed Dr Rashmi Mayur from the International Institute for Sustainable Future in Mumbai, India for a film about his life. Ross and I had met him at the Social Summit in 1995 in Copenhagen and developed a close friendship and cooperation with him. We saw our relationship with Rashmi as an assurance and control of whether our thinking was truly global. He was part of founding GEN and we remained in close contact until his death in 2004.

I asked Rashmi what he saw as the purpose of his life. He answered immediately in his characteristic up front style as if this was always present in his mind, "Number one: I want to help the poor of the global South create global justice. Number two: I really want to know: Why are we here? Who are we? What is the purpose? Number three: More than anything, I want to save three billion years of evolution on Earth." Those sentences have been resonating within me ever since. I shall adopt his points as a starting point for two articles: here in 'Living the New Worldview' (outer worldview) and in 'Who am I?' (inner worldview) in module two. I believe that Rashmi had a clear and short way of defining the essence of living the new worldview on multiple levels.

The Need for Global Justice

The new worldview teaches we are all one undivided humanity. We are interconnected. We cannot hurt others or nature without hurting ourselves. Spiritual leaders in many traditions across the globe proclaim these timeless truths, yet it seems that relatively few of us have adopted these lofty teachings in practice, even if we believe in them or give lip service to them as a new worldview.

What does the Oneness of humanity mean in practice? It means that all people all over the world have the same rights to a good life as I do, coming from the privileged North/West. It means that the current inequities – with the global North taking more than its fair share of resources, energy and material riches – are unfair not only to the Global South, but are harmful to all of us. If all people were to live as we do in the North, humanity would requite five planet Earths! It also means that an economic system preventing countries in the South from developing must be changed.

The global South rightfully feels frustrated and angry with the Northern resistance to creating global justice. Many, especially young people, are suffering from mass unemployment, degradation of the environment, hopelessness, no belief in a future and contagious diseases. If we do not help bring about real justice then we will see the result on our own doorstep: refugees, migration, drugs, diseases, wars, natural disasters, boycotts, and collapse.

First and foremost we need to see every single person as a human being – not as a Moslem, a Jew, a Hindu, a poor person or a sick person. We must see each individual as a divine being. In India, when receiving guests or marrying, a garland of flowers is placed around the neck signifying: I adore the divine in you. This is the proper attitude. If we *truly* adopted this attitude of deep respect along with it would come a natural willingness to relinquish certain acquired rights and excessive material consumption. It would entail a willing reduction of the 'ecological footprint' of the North by 80% so that the South can have its proper share.

First and foremost we need to see every single person as a human being – not as a Moslem, a Jew, a Hindu, a poor person or a sick person. We must see each individual as a divine being.

This process is difficult as it requires a revolution in the world economy, a shift in the political systems of many countries, a shift in consciousness of unequalled dimensions and a huge program of education for sustainability. Admittedly this is no small task but it is not unfeasible; we are talking about human choices here, not inevitable outside forces. One way to advance equal rights for all planetary citizens, as German Chancellor Angela Merkel suggested in 2006, would be to adopt the equal right of everyone to the same CO_2 emission. There are other reasons for global imbalances, based in cultural patterns and religions, which are beyond our capacity to influence. What we can do is to sweep in front of our own door.

David Korten has provided an excellent analysis in his book *The Great Turning, From Empire to Earth Community*. Empire is the attitude of domination and Earth community is one of partnership and cooperation. David Korten identifies the neo-liberal economic system and the WTO as being chiefly responsible for growing inequalities across the globe. These issues are discussed in greater detail in *Occupy Worldstreet* by Ross Jackson and in *Gaian Economics*.

Much has been happening globally since the reality of climate change and peak oil have been largely accepted. Al Gore's film *An Inconvenient Truth* and more than 1000 of his talks around the globe have been convincing many people. The election of Barack Obama as president of the US gave new hope for a short time. Global leaders failed to agree to any serious reductions of CO_2 levels at COP 15, 16 and 17. Financial crises are tearing countries apart with no solution in sight. No wonder that the global commons are going to the streets with the Occupy movement.

Saving 3 billion Years of Evolution

30 years after introducing the Gaia hypothesis, James Lovelock warned us that Gaia is losing her capacity to restore ecosystems and that evolution itself is under siege. The extinction of many of our plants and animals is threatened if four degrees of global warming occurs. There is also the possibility that if we die, Gaia, our living planet, will also die. We are like in a vessel just before the Niagara waterfall with a broken down motor. Neither the free market, globalisation nor the sustainability movement is enough to make the radical changes needed, in Lovelock's opinion. He proposes nuclear power as a shot-term solution to reduce atmospheric CO_2.

When Lovelock argues that the sustainability movement is not enough, I have to agree with him. He is not, however, aware of ecovillages as a solution. Nor the concept of Earth Community which is so much more radical as a social solution than the sustainability movement he describes. Nor does Lovelock have the spiritual perspective. The transformation of consciousness can change the goals of humankind from materialism to wanting a simple life. We need to take this seriously. It is our primary responsibility to save Gaia. For many, nuclear power has lost its attraction after the Japanese tsunami and earthquakes in 2011. The next step is the realization that changes have to be on a very fundamental level. We need to go deep.

In his book, *The Dream of the Earth*, Thomas Berry has given us the necessary framework for these changes. It is relevant for all religions and creation stories around the planet (see pages 52-57). The Universe Story is teaching us the role of Gaia in the Universe and the necessary language and story to think in new ways about her and us. "We need to reinvent ourselves as humans," as he writes. Ecovillages fit well into this story. A simple, local life in sustainable abundance not using more than our share of the allowed CO_2 is possible. Living a worldview of Oneness.

Berry's second point is equally important: "We are not a collection of objects but a communion of subjects." All too long politicians and corporations have treated people like objects who should simply accept life as it is, including a financial system that ignores the wellbeing of the '99%', as the Occupy Movement term it. But people are waking up all over the planet as evidenced by the Arab spring of 2011. We want to be subjects, not objects, creating a life we want.

Elisabet Sahtouris, an evolutionary biologist, has provided further language and a knowledge to enables us to understand how we are part of a Living Planet, where cooperation is as important as competition (see pages 18-24). She explains how we may become co-creators of evolution and redefines science not as a science of death, but as a science of life on a living planet. All her insights, materials and books are available for free on the Internet (see www.sahtouris.com).

My primary response to Rashmi's concern for preserving three billion years of evolution is living these new insights in the ecovillage movement. This movement, when taken as a whole, offers inspiring and practical examples of sustainable living and hope to the world.

Implicit is the necessity of change driven by the grassroots. We need an unprecedented global social movement like the World Social Forum from 2001 plus the Occupy Movement to move us from an Age of Excess to an Age of Moderation. Many communities have now taken the lead. Just as ecovillages and Earth Democracy are democratic solutions from below, the Transition Movement concept is gaining momentum all over the world. Can we quickly transform our consciousness and patterns of habitation towards sustainable living and thereby reduce energy consumption on a massive scale? I believe so. It would be no more difficult than building nuclear plants all over the planet and then trying to cope with the unparalleled storage and security risks arising from this industry.

The Global Change of Consciousness

The need for both inner and outer change is being recognized widely. Therapies and contemplative work has been flourishing for 30-40 years, slowly changing people's attitudes. The number of 'cultural creatives' has risen to 30% of the population in the USA and Europe. Eastern teachers have, for more than 100 years, been bringing the ancient wisdom to the West. Scientists and spiritual practitioners are meeting in practical organisational ways.

Ken Wilber has offered an Integral practice for many people and created research institutes in all fields of knowledge in the USA, influencing both seekers and scientists. The Naropa Institute has built a university on spiritual Buddhist values. Living Routes has been bringing student from the USA to ecovillages worldwide to learn. The GPIW, Global Peace Initiative of Women are arranging meetings all over the globe by faith leaders so they start cooperating. Andrew Harvey is creating *Grace Groups* (where people let their hearts break open and together they define how they can act together) and is linking them all over the world.

Recently some exciting Internet universities have emerged having the same purpose of uniting science and spirituality and reaching across the globe. The Wisdom University was initiated by Jim Garrison as a way of spreading the knowledge originating from the meetings 'State of the World Forum' (in the 1990s). The Ballaton group from Hungary has created the Giordano Bruno University (Bruno was burnt on the stake in 1600 for wanting to create a religion of love) with the aim of bringing cheap Internet education to all the world with Erwin Lasslo as its head. Gaia Education and Gaia University are other global initiatives.

Ecovillages as a Global Just Solution

GEN was founded in 1995 with a strong motivation to facilitate the transformation to sustainable communities and societies by recognizing the need to begin by putting our own house in order. This is a modest beginning to building an equitable basis for cooperation amongst all peoples. For 15 years GEN has been building networks all over the world, publishing books and gathering experience so that knowledge about sustainable lifestyles is now available for the world to tap into. In 2005, a new educational program, Gaia Education, was launched to pass on the practical knowledge gained from the ecovillage movement.

For me, the ecovillage idea is a way of combining the three goals that Rashmi expressed: both the inner and outer change of worldview. The ecovillage concept provides us with a global vision, possible for all, and could create a just and sustainable society if only the global economic system and politicians would be supportive of this model. We could learn to live within our limits, on one planet instead of five, if we all adopted this lifestyle. We could reduce the global North's footprint by 80% and leave room for the South 'to develop'. This would be a simpler life but with higher quality.

In many places, the global South has accepted the ecovillage idea immediately because it builds on existing village structure and spiritual traditions. The Northern ecovillages of GEN have learnt much from Sri Lanka, Ladakh, Africa and from Asia's engaged spirituality movements. As Satish Kumar describes in his brilliant article 'The Spiritual Imperative' (see pages 202-213) what a spiritual life is and that simplicity itself is a goal.

David Korten's book *The Great Turning: From Empire to Earth Community* has created a mainstream context for action. He advocates 'Earth Community' as the solution globally. Vandana Shiva speaks of 'Earth Democracy'. The international lawyer, Polly Higgins, is campaigning for the Earth's Planetary Rights. Ecovillages are already existing Earth Communities. They are full expressions of conscious choices about how we want to live in partnership and in community with our fellow human beings and with plants and animals.

Ecovillages are systemic solutions to the world's energy shortages and the need for a simpler life. Transportation is reduced as is the need for multiple heated spaces. Footprint analyses demonstrate that the footprint of people living in ecovillages is reduced to less than 40% of the UK national average.

Ecovillages often function as earth restoration projects. They cherish diversity and maintain areas for wildlife and for nature sanctuaries. When groups of people live on the land and together make decisions for the land, they listen, respect and honour it. Business interests are reined into balance with conserving biodiversity and natural resources. Local economies are reinvented. Ecovillages and the concept of Earth Community are compatible with humans becoming co-creators of evolution and thus save and continue three billion years of evolution.

Will humanity start building an ecovillage future? This will depend on our spiritual insights, the media and politicians, and the outer threat of natural disasters. Rashmi firmly believed in ecovillages and Earth Community. He is not alone.

Hildur Jackson, born 1942 is trained in law and cultural sociology, permaculture design and spiritual development. She is a long-time grassroots activist and initiator, with her husband of 45 years, Ross Jackson, of one of the first Danish cohousings in 1972. She is a co-founder of Gaia Trust in 1987, Denmark and GEN (The Global Ecovillage Network) in 1996 and Gaia Education in 2004. She lives on the Farm Duemosegård. She has three grown sons and soon seven grandchildren. She has edited *Ecovillage Living: Restoring the Earth and Her People* (Green Books, UK, September 2002), with Karen Svensson and *Creating Harmony: Conflict Resolution in Community* (Permanent Publications, UK, 199?) which shows how people in communities all over the world have invented ways of living together peacefully. See **www.gaia.org**

A meditation that explores how within the human heart we can discover that the vastness of the cosmos and the great miracle of life is held within our true Self.

Guided Meditation: Human Being as Miniature Galaxy

William Keepin, PhD

This meditation begins with a passage from one of the timeless scriptures of India, the Chandogya Upanishad:

> *In the center of your heart, no bigger than the size of your thumb, is a secret dwelling, the lotus of the heart. Within this dwelling is a space, and within that space is the fulfilment of all desires. As great as the infinite space beyond is the space within the lotus of the heart. Both heaven and earth are contained in the inner space, both fire and air, sun and moon, lightning and stars. Whether we know it in this world or know it not, everything is contained in that inner space.*
>
> *Never fear that old age will invade that space; never fear that this inner treasure of all reality will wither and decay. This knows no age when the body ages, this knows no dying when the body dies. This is your true Self, free from old age, from death and grief, hunger and thirst. In this Self, all desires are fulfilled.*

Imagine now this space in your heart, about than the size of your thumb. Center your attention there, and bring all of your awareness to focus on this center. Now imagine that this space in your heart is a gateway to the infinite, to the vast cosmos beyond, as the Upanishad says... As you focus inward, picture in your mind the night sky, with its billions of galaxies... each one having a billion stars. As you gaze upon the night sky, looking from one horizon to the other, you see the vast expanse of the universe. Stars are twinkling from afar, each one is sending light rays, or photons, to your eyes. These photons arriving in your eye come from different stars, some of which are spaced millions of light years apart from one another. So the vast distances of the universe are replicated, in miniature, right inside your eyeball. And not only this vast spatial expanse, but also the vast temporal history of the universe is replicated inside the tiny space of your eye. What you see when you look out upon the night sky is not the universe as it is now, but as it was in ancient times, when those photons were first created.

Consider the lifetime of one of these photons arriving in your eye from a distant supernova star located some 70 million light years away. That little photon was first born 70 million years ago. When it came into existence, it immediately began racing through space at 300,000 kilometers per second. This speed is fast enough to zip around the equator of the Earth seven times in less than a second. At this breakneck speed, this little photon started its journey, back at the end of the Mesozoic era, when the Dinosaurs were proudly lumbering upon the earth. As this photon kept racing toward the Earth over the next five million years, the dinosaurs went extinct, probably caused by a meteor that crashed into the Earth. Then mammals began to evolve, and in 5 million years, squirrels, herons, and storks appeared. It took another 5 million years for rabbits to appear, and another 5 million for monkeys. All the while this little photon continued scurrying its way across space. Another 30 million years, and the first chimpanzee walked the earth. Another 8 million years, and the first bipedal hominids appeared. As our little photon kept faithfully racing its way toward earth, another twelve million years passed, and the early Neanderthal *homo sapiens* appeared, the first prototype humans. Then another 200,000 years and the early Altamira cave paintings were done, and after another 20,000 years, writing was first developed in Sumeria. Now our little photon is nearing the end of its long journey, being a mere 6,000 light-years away from its destination. In these last few moments of the photon's vast journey, we witness the rise and fall of all the ancient human civilizations, and then move right up to the last couple hundred light years, with the rise of modern industrial civilization. And then, when the photon is just tens of light-years from its destination, you were born. Throughout your entire childhood, youth, and adulthood, this little photon kept racing toward you. Eventually, that special moment comes one night, when you step out on the deck to enjoy the stars. As you look up, that little photon finally enters the pupil of your eye, travels the last two centimetres of its journey, and then gives up its life on your retina. This tiny, luminous being – born when the dinosaurs roamed the earth, then traveling at the speed of light for 70 million years – now dies, in order that you may see. It becomes transformed into a tiny electrical pulse in your optic nerve, which relays to your brain a direct visual report from that particular supernova all those eons ago.

Billions of similar photons are arriving in your eye at the same moment, coming from across the expanse of the universe, each one bringing you a visual history of its star of origin. Taken together, these photons provide a luminous miniature replica, within your eyeball, of the vast reaches of space and the ancient history of the universe. All this takes place within the tiny space of your eyeball, as you gaze upon the night sky...

Now let us follow the path of our little photon further after it enters your eyeball. It becomes an electrical impulse, which then moves to your brain, and is absorbed there as energy into your body. But what is this body?

Let's focus more closely on this mystery that is your body. What is its structure? Your body is comprised of five billion billion billion atoms (5×10^{27}). This is a lot of atoms, but just how big a number is this?

Imagine that each atom were the size of a pea, the little green food that we love to eat. How big a pile of peas would we need to equal the number of atoms in your body? Suppose we filled your entire house, floor to ceiling, with peas. Would that be enough peas? Not even close. Suppose we filled every building with peas in a big city, like New York. Do we then have enough peas to equal the number of atoms in your body? Again, nowhere near. OK, let's say we covered the entire nation of [whichever country the meditation is being conducted] with peas, right up to your neck. Would this be enough? Still not even close. Alright, now let us cover the entire earth – land and sea – up to our necks in peas. That's 1.5 meters (or 5 feet) deep in peas, across the entire surface of the earth. Surely this would be more than enough peas? Again, not even close. It turns out we would require *one million* planets, each one the same size as the Earth, each of these planets covered neck-deep in peas... in order to equal the number of atoms your body.

Now consider just one of those atoms, and let's zoom in on it. Imagine a single atom in the tip of your finger, right now, as you sit there. Imagine that you can zoom in on this atom in your finger. We are zooming in now, and we magnify it ten trillion times. This atom now becomes about the size of a soccer field. On this soccer field, the electrons are sitting on the goal lines. Each electron is about the size of a soccer ball. And right in the middle of the soccer field, if we look very carefully in the grass, we can just barely find the nucleus of the atom. It's the size of a tiny pinhead. Even though this nucleus weighs at least 1,800 times more than the electron, it's much smaller. And all the rest of the atom is empty space. Nothingness. All matter is comprised almost entirely of empty space. As the Buddhist Heart Sutra says: "Form is emptiness. Emptiness is form." The Buddhists call this emptiness *shunyata*.

What this means is that you, sitting there right now... are almost entirely empty space. You are mostly emptiness. If you squeezed out all the gaps in your body, like squeezing out the holes in a sponge, you would shrink down to a tiny spec of dust so small that you could barely see it with a magnifying glass. It would be a heavy little spec, having the same weight as you do now. But that tiny spec is the totality of all the physical material that is in you. The rest of you is empty space. All five billion billion billion atoms in your body would fit inside that tiny spec of dust, if you squeezed out all the gaps.

But what is an atom made of? Atoms are made of light, sometimes called 'frozen light', because atoms are made of light that has been solidified into matter. So your body consists of 5 billion billion billion tiny glistening jewels of light, separated from each other by vast expanses of empty space.

All matter is comprised almost entirely of empty space. As the Buddhist Heart Sutra says: "Form is emptiness. Emptiness is form."

You are literally a miniature galaxy – with billions upon billions of points of light, just like tiny stars, all held together in a vast empty inner space. Your atoms are clustered in various different ways, and they swirl about here and there in organized structures, like miniature solar systems.

So you are something like a child's dot figure – a huge array of glistening points of light, separated by vast empty spaces. Only when you connect the dots does the form appear, and you come alive. And what connects these dots? What holds it all together? Love. In science, these connecting forces are given various technical names – electromagnetic, gravitational, strong and weak nuclear forces – but in truth, these are just so many forms of love. You are made of tiny points of light, all woven together by love in a perfect tapestry. This love is the same love that holds the planets and galaxies together.

You are a galaxy of light, knit together by love. And you are in turn held in the embrace of love by the Earth… which is held in the embrace of love by the sun… which is held in the embrace of love by the galaxy… which is held in the embrace of love by the universe… which is held in the embrace of love by the ineffable infinite (God)… There is a Sufi saying, "I was a hidden treasure, and longed to be loved, and so I created the universe." Feel this treasure now, that is you… Feel yourself as the unique and precious treasure that you are… one with the treasure of the universe…

The secret song in your heart… is the entire universe singing…

The galaxy that is you… and the galaxies that comprise the universe… are a single living treasure… in the heart of the Divine.

> *By love has appeared everything that exists.*
> *By love, that which does not exist appears as existing.*
> Shabestari

…Taking all the time you need, gently begin bringing your awareness back into 'ordinary reality', in this present moment, and allow your eyes to gently open, remembering this glorious secret of who you are.

Please see page vi for William Keepin's biography.

MODULE 2
The Awakening & Transformation of Consciousness

Contents

Earth As Sacred Community

The Urgent Need for Spiritual Awakening

Spiritual Responsibility in a Time of Global Crisis

Living in Auroville: a Laboratory of Evolution

Who Am I? Why Am I Here?

The Great Transformation

The Great Turning

The Shambhala Warrior

"A human being is part of the whole – called by us the Universe, a part limited in time and space. He experiences himself, his thoughts and feelings, as something separate from the rest, a kind of optical delusion of his consciousness. This delusion is a kind of prison for us, restricting us to our personal desires and to affection for a few persons nearest to us. Our task must be to free ourselves from this prison by widening our circle of compassion to embrace all living creatures and the whole of nature."

Albert Einstein

Thomas Berry explains why the new story of the universe is a biospiritual story as well as a galactic story and an Earth story. This story is one of integration and interdependence, providing a new view of the Earth that is at once scientific, spiritual, and shimmering with life and peace.

Earth as Sacred Community

Thomas Berry

In our discussion of sacred community, we need to understand that in all our activities the Earth is primary, the human is derivative. The Earth is our primary community. Indeed, all particular modes of Earthly being exist by virtue of their role within this community.

Failing to recognise this basic relationship, industrial society seeks to subordinate the entire Earth to its own concerns, with little regard for the consequences for the integrity of the planet. This subjugation of the primary functions of the Earth to the limited concerns of the human can be observed in all our professions and institutions.

In a multitude of different ways, we seek to subdue the Earth to our own ephemeral purposes, considering this the proper human relationship to the natural world. Because of this distortion in our thinking, we are carrying out what may be one of the most devastating assaults ever on the Earth in more than four billion years of life on this planet.

Something much greater is happening than is generally realized. This is something more than the 'end of the Enlightenment' as is sometimes suggested. It is something more than a parallel with the fall of Rome. It is something more than the demise of Western civilization. It is a transition greater than any historical transition that has ever taken place in human affairs. For this is a deleterious change not simply in human social order or cultural expression. It is a change in the very chemistry of the planet that makes life possible. It is a change in the biosystems of the planet. It is a change in the geological structure of the planet.

Not simply the human future is involved. The future of every living being on the planet is at issue. The fate of the planet itself in its most profound physical and psychic structure is being determined. We are witnessing nothing less than the dissolution of the planet Earth and all its living systems in consequence of this strange distortion of our human role in the Earth process that has emerged from within our modern Western world, which was itself born out of a biblical-classical matrix.

Here we might observe that our Western religious institutions are strangely indifferent to what is happening. This indifference arises, apparently, as a result of excessive concern for redemptive processes out of this world – which is considered to be seductive – rather than integration within this world considered to be sacred. There seems to be little realization that the disintegration of the natural world is the destruction of the primordial self-manifestation of the divine. The very existence of religion is threatened in proportion as the splendor of the natural world is diminished. We have a magnificent sense of the divine because we live in such a resplendent world. If we lived on the moon, our sense of the divine would be as dull as the lunar landscape.

Even when we try to bring religious influence to bear on these issues, we find that our religious traditions have little relevance to what is happening. Our Western religions exist in a different world, a world of covenant relations with the divine, a world little concerned with the natural environment or with the Earth community. Our sacred community is seen primarily as one concerned with human-divine relations, with little attraction toward a shared community existence within the larger world of the living. Our iconoclasm is such that we can hardly think of ourselves within a multispecies community or consider that this community of the natural world is the primary locus for the meeting of the divine and the human.

One study done at Yale University found that the more extensively people participate in religious activities, the less likely they are to be concerned with the natural world. The pathos of the human, described so extensively in the prophetic writings, seems to exhaust our religious energies. Religious attention is directed toward moral conduct, social injustice, pietistic practices, and interior meditation experiences.

Valid as these activities might be, they are themselves frustrated at present because the primary reorientation of our society toward a more integral relation with the Earth is not taking place. Religion, rather than supplying the needed cultural therapy, is in need of a profound rethinking of itself and its role in Earthly affairs. This requires that we humans reflect on our present situation to understand how it has come about, why we are so incompetent in our efforts to mitigate the damage being done, and how we might foster a profound Earth renewal.

The continuity of the human with the natural world in a single sacred community can be appreciated in the experience of Black Elk, a Lakota Indian. When he was nine years old, he experienced an elaborate vision culminating in a vast cosmic dance evoked by the song of the black stallion seen in the heavens: "There was nothing that did not hear, and it was more beautiful than anything can be. It was so beautiful that nothing anywhere could keep from dancing. The virgins danced, and all the circled horses. The leaves on the trees, the grasses on the hills and in the valleys, the waters in the creeks and in the rivers and the lakes, the four-legged and the two-legged and the wings of the air – all danced together to the music of the stallion's song."†

This continuity between the human community and the natural world was altered by identifying the human as a spiritual being in contrast to all other beings. Only the human really belonged to the sacred community of the redeemed. The previous sense of a multi-species community was diminished.

The humanist traditions that come down to us through the classical writings of the Greco-Roman world, apart from those of the Stoics, supported this alienation. This is evident in their emphasis on human grandeur considered to be distinct from the wilderness world and opposed to any sense of multispecies community. This 'arrogance of humanism', as David Ehrenfeld termed it, alienated Western society from both its religious heritage and its intimacy with the natural world. Whatever the human gains in these Western religious and cultural developments, the Earth in its most essential functioning has been profoundly disturbed by these developments.

As regards our own specifically Western responsibilities, we must note that, although we have developed a moral teaching concerned with suicide, homicide, and genocide, we have developed no effective teachings concerned with biocide, the killing of the life systems of the Earth, or genocide, the killing of the Earth itself.

Realizing that there is something terribly wrong in our relations with the natural world, our religious traditions have recently begun putting special emphasis on the concept of stewardship as the primary relationship between the human community and the natural world. This concept of stewardship is derived from biblical statements concerning human dominion over the Earth and all its living creatures. To many religious people, this seems quite adequate as a basic orientation toward the natural world. To others, stewardship itself is the origin of our present evils. There is no way in which we can care for the natural world or improve on the genius of nature. It would be difficult to discover any human improvements that have ultimately been beneficial to the natural life systems of the Earth, although we can on occasion bring about a measure of healing to the damage that we have caused.

Yet it should not be said that stewardship exhausts Christian concern for the world about us. There is a profound Christian awareness that the natural world is itself a manifestation of the divine. This has led to the concept of revelation being contained in two scriptures; one the scriptures of the natural world, the other the scriptures of the Bible. While this sense of the natural world as revelatory has been severely diminished since the sixteenth century, with the discovery of printing and the consequent emphasis on the written word, it is still available in the tradition itself. If more fully developed, this could lead to a more effective concern for the survival of the planet.

The difficulty with the story of the universe, as this is known to us through recent centuries of intense observation and analysis, is that it has generally been told in an inadequate way, as simply physical in form and random in

> *There is a profound Christian awareness that the natural world is itself a manifestation of the divine.*

process. Insofar as it goes, this scientific presentation of the universe has achieved a certain dazzling success. Yet the materialistic interpretation generally given to the scientific data has become progressively less adequate.

The entire scientific process was severely affected by Werner Heisenberg's discovery in the 1920s that the reality known is profoundly affected by the knowing subject, that our knowledge of the universe never ultimately attains the so-called objective world in any absolute sense. Knowing is a communion of subjects rather than a simple subject-object relationship.

Several other considerations have also affected our overly facile scientific presentation of the universe. In such presentations there is no reference to the fact that all coding, whether in the structure of the elements or in the genetic endowment of living beings, is by its nature an immaterial, a psychic determination. While certainly not separate from the physical aspect of things, this inner psychic dimension is distinct from any of the component parts as well as from the totality of the parts. It is the unifying principle imperceptible to sense faculties but immediately known by intelligence, just as a tree is seen as a unity in all its physical relationships and is not reducible to any or all of these. There is a mystical dimension not only in the reality known but in the scientific equations whereby we give expression to the knowing act. Our study of genetics may provide us with a pattern of the components governing the genetic process, but we will never have physical experience of the inner principle that enables this amazing complex of genetic materials to function with the unity and spontaneity we observe in the unfolding processes of life.

Upward Integration

A further observation: if we proceed in scientific inquiry by a downward reduction of wholes to their parts, we need to complement this procedure by an upward integration, in which we understand parts by their function in the larger configurations. We cannot understand any part of the universe until we understand how it functions in the whole. For example, any study of the element carbon limited to its inanimate form provides only minimal understanding of carbon, since carbon has astounding capacities for integrating the basic elements needed for organic existence. Even beyond the organic and the qualities associated with living beings, there is the capacity of carbon to enter into the processes of thinking. Thought itself and the highest of human spiritual achievements are attained through activation of the inner capacities of carbon and its alliance with the other elements of the universe. Thus carbon has varied modes of expression, from inorganic to organic to conscious self-awareness in the human.

In the interconnection, then, of part to whole (as seen in the example of carbon), there is surely a certain discontinuity that emerges, but also a continuity that must not be neglected between the different modes of expression – inorganic, organic, reflexive. The materialism of science or the spiritualizing tendencies of religion that refuse this continuity of the human and all our capacities with the natural world ends up with a radical disassociation of the human from the universe about us. Moreover, to identify this disassociation with spirituality is to mistake the entire meaning and significance of spirituality in its human expression.

Vast periods of time have been required to bring about this sequence of transformations from the primordial radiation to the shaping of the elements, from the elements to the molecules, then to the megamolecules – the viruses, cells and organisms – and on to the more complex living forms, and eventually to the human mode of consciousness.

Narration of this sequence has required the immense effort of scientific investigation of these past few centuries. It has necessitated the setting aside, for a while, of the spiritual, visionary, intuitive, imaginative world in order to probe as deeply as possible into the visible, material, quantitative world, the measurable world, the world that could be expressed in the language of calculus, the great instrument of the scientific endeavor. The success of this scientific achievement and its subservience to the often-ephemeral purposes of the commercial, industrial and financial establishments produced a profound revulsion in many religious persons. Out of this revulsion has come a widespread reassertion of traditional religious teachings with a fundamentalist fixation.

... antagonism between mechanistic science and fundamentalist religion is one of the basic reasons for our inability to establish a sense of sacred community with the natural world.

This antagonism between mechanistic science and fundamentalist religion is one of the basic reasons for our inability to establish a sense of sacred community with the natural world. Scientific materialism could not evoke the awe or wonder or fear of the natural phenomena needed to restrain the more grasping tendencies of the human. Nor could fundamentalist Christianity committed to redemption out of this world evoke the dedication needed for intimacy with the forces of nature, considered to be both seductive and without spiritual meaning. Strangely enough, even some of the other religions, such as those in Asia, with magnificent traditions of intimacy with the natural world, could not restrain their own adherents from plundering the Earth in the race to consume more resources for modernization and development.

My proposal is that we cannot fully remedy this situation except by a realization that the universe from the beginning has been a psychic-spiritual, as well as a physical-material, reality. Within this context the human activates one of the deepest dimensions of the universe and is, thus, integral with the universe from its beginning. The universe story needs to be accepted simultaneously as the human story and the story of every being in the universe.

There is a need for the religious traditions, on their part, to appreciate that the primary sacred community is the universe itself, and that every other community becomes sacred by participation in this primary community. The story of the universe is the new sacred story. The Genesis story, however valid in its basic teaching, is no longer adequate for our spiritual needs. We cannot renew the world through the Genesis story; at the same time, we cannot renew the world without including the Genesis story and all those creation stories that have nourished the various segments of the human community through the centuries. These belong to the great story, the sacred story, as we presently know this sacred community.

The New Story of the Universe

The New Story of the universe is a biospiritual story as well as a galactic story and an Earth story. Above all, the universe as we now know it is integral with itself throughout its vast extent in space and throughout the long series of its transformations in time. Everywhere, at all times, and in each of its particular manifestations, the universe is present to itself. Each atomic element is immediately influencing and being influenced by every other atom of the universe. Nothing can ever be separated from anything else. The Earth is a single if highly differentiated community. This is the quintessential way of understanding the universe.

So too, every part of the universe activates a particular dimension or aspect of the universe in a unique and unrepeatable manner. Thus everything is needed. Without the perfection of each part, something is lacking from the whole. Each particular being in the universe is needed by the entire universe. With this understanding of our profound kinship with all life, we can establish the basis for a flourishing Earth community.

† John G. Neihardt, *Black Elk Speaks,* Lincoln, NE, Univ. Nebraska Press, 2004, p. 32.

Thomas Berry died on 1 June 2009. He was a leading visionary thinker on the foundation of human cultures and their relations with the natural world. A self-described 'geologian', his work uniquely integrates spirituality and ecology while transcending the limitations of both. Berry's most important books include *The Dream of the Earth* (1988) and *The Universe Story: From the Primordial Flaring Forth to the Ecozoic Era* (1992) written with cosmologist Brian Swimme and *The Great Work: Our Way into the Future* (1999), Bell Tower/Random House, NY.

Buddhist nun, Jetsunma Tenzin Palmo, explains why spiritual liberation will not only transform individual consciousness, it will transform the state of the planet for the better. We don't achieve lasting fulfillment by pursuing our desires, but by giving joy to others. Others have done it, and we can too.

The Urgent Need for Spiritual Awakening

Jetsunma Tenzin Palmo

*The word 'Buddha' means to awaken.
We are all asleep, we are all dreaming,
and we believe our dreams.
This is the problem.*

One time when I was living in the cave I had a dream. I dreamt that I was in an enormous prison without end. In this prison, there were many levels. There were the penthouse suites where people were laughing and talking and dancing and making love and working. There were the levels all the way down, until you got to the dungeons where people were writhing in agony and despair of mind. But whether we were in the penthouse or in the dungeons, we were all in the prison. I suddenly realized it was so insecure; people in the penthouse today could be in the dungeons tomorrow. We were all trapped together. We had to get out. So I spoke to a number of my friends: "Look, this is a prison, we have got to leave." They all said, "Oh yes, it's a prison, but it's okay, it's not bad." Or they said, "True, it's a prison, but it's so difficult to get out. It's better to accept the fact that we are here." Eventually I found two friends who agreed to try to escape with me, and the dream went on.

    The question is, why do we regard our ordinary life here as a prison, and how do we get out? This is basically the question in Buddhism. But why is it a prison? You might say, "My life is okay, it's not a prison. I can more or less do what I want to do." It is not dealing with the physicality; it is dealing with the mind. Our minds are imprisoned, not by external gates, but by ignorance. This is so universal, and it is why I am troubled for the future. Despite all our external learning, our research and science, we are still absolutely ignorant. Spiritually, we are as ignorant as we were when the Buddha walked this earth. What are we ignorant of? Einstein said that in our age there has been a tremendous growth in knowledge, but absolutely no growth in wisdom. Ignorance has nothing to do with education, nothing to do with external brilliance and genius of mind. What are we ignorant of? We are ignorant of our true being and what is really the nature of this world. Because we are clinging to all the wrong things governed by ignorance, we are enslaved.

Buddhism is always concerned with how to become free. It is always concerned with liberation, liberation of mind. The problem is that normally we live within a world of time – past, present, future – and in a subject/object dichotomy. There is the subject 'me' and the object 'everyone else out there'. We cling to this sense of 'me' and 'mine-ness'. Some people think 'I' when they think of their gender, their race, their country, or their religion. They think, "This is who I am. I am my personality. I am my memories. I am the sum of all these things."

Some people are more subtle, and they say, "No, behind all that there is something else. There is an 'I' which is unchanging, which has always been there since I was born up to now." But when you look to find this 'I' which separates 'me' from all the 'you' out there, where is it?

Buddhism is not just to make us calm and quiet and feeling happy. It is to peel off the layers of our onion of individuality. If you peel off the various layers, the first layer race, then the layer gender, then nationality, then education, then one's level in society, one's profession, where is this 'I'? Eventually you get to something else which is totally beyond 'I'. This intrinsic awareness, this primordial awareness, which is at the very basis of our being, has nothing to do with 'me' or 'you'.

We experience a level of awareness behind the coming and going of thoughts and feelings and concepts. It is a wordless, timeless, non-dualistic perception. If we can remain always in that higher level of total awareness, we are Buddha. It is simple. This awareness is not something up there, and it is not actually something that is difficult to realize. Awareness is just awareness. The Tibetans compare it to the sky. The sky has no centre, and it has no circumference. It is endless. The sky is not just up there. It is within and all around us. It is space. In Tibetan, the word for space and the word for sky is the same word. So where is space not? Where is this awareness not?

The word 'Buddha' means to awaken. We are all asleep, we are all dreaming, and we believe our dreams. This is the problem. When we awaken even for a moment, then we see that what we cling to is really our own projection. Then our minds are so sharp, so clear, and so awake, and we realize that our true nature is something completely beyond the conceptual thinking mind. The important thing is that externally nothing changes, but inwardly everything changes. Everything becomes alive and clear and vivid, but there is no ego driving it. Then everything spontaneously happens, whatever one needs to do is spontaneously accomplished, without the ego

The word 'Buddha' means to awaken. We are all asleep, we are all dreaming, and we believe our dreams.

Einstein said that in our age there has been a tremendous growth in knowledge, but absolutely no growth in wisdom.

getting in the way. It is accomplished skilfully.

What stands in the way of our liberated mind? This is what we have to deal with and what is happening in our society nowadays. What stands in the way of our realizing our minds? The true nature of our mind is obscured like a thick cloud that covers the blue sky. That cloud is made up of our negative emotions, like our clinging greedy minds, our anger and aversion, our pride and arrogance, our jealousy and envy, and especially our ignorance of not realizing our true nature. And this acts like a screen. Do we realize how much we live our lives through our minds? Everything we see, everything we say, everything we do, is directed by our minds, our thoughts, our feelings, our memories, our concepts, our judgments.

We hardly see anything as it is. We see our opinion. It is very hard to see things nakedly without the many sheaths of our conceptual opinions and ideas about that thing. We come here and we look at this ceiling. Either we think this is magnificent art, or we think it is absolute kitsch. We think it is wonderful, or we think, "Oh my God, how could anybody have done this?" It makes no difference; the ceiling is just a ceiling and the painting is just paint. How we react to it depends on our mental framework, our background, our education, our aesthetic taste. Everything is like that. We never see things as they really are, we only see our version. Everything we experience, we experience through our mind. Everything we see, we hear, we taste, touch, or feel, is interpreted through our mind. Yet we do not know the mind itself.

We say, "I think that, I feel that, in my opinion it is that." But what is a thought? What is a feeling? What is an opinion? We are always streaming outside through our senses, but we never turn that awareness, which sees and thinks and tastes and touches, inward onto the mind itself. What is a thought? Where does it come from? What does it look like? Where does it go? And who is thinking? If we say, "I am thinking!" Who am I? What is this whole thinking process and what is behind the thinking process?

We are so caught up in our heads. Some neurologists say that nowadays we know so much about the brain, but we still have not found the mind. In Asia the mind is not up in the brain. The brain is the computer, but the source of the mind is somewhere down here (in the centre of the chest). It is very interesting that when you first start meditating, you are meditating in the head. There is the mind thinking and the meditation practice you are trying to do. So it is like they are both facing each other. It is you and the practice. This dualistic approach which we start with is up here (in the head). The brain is trying to meditate. Once the meditation really kicks in and the mind really goes into a state of meditation, the meditation itself goes down to here (in the centre of the chest). Then there is no meditator and no meditation. You become one with the practice. At that time, things start moving. This is something you experience. It is not something you think about. As long as you are thinking about it, it stays up in the head. When you become the meditation, it moves down to the centre of the chest, as all religions have always known.

In particular, what troubles us in our modern culture? The Buddha said that the causes of our suffering are our negative emotions, especially our ignorant clinging to an ego. Our greed means "I want" for this ego, and our anger and hatred means 'I don't want' for this ego. This is the cause of our suffering.

Our modern society is selling us the idea that if we could only fulfil our desires, we would be happy. Two thousand years ago, Buddha said "Desires are like salty water. The more you drink, the thirstier you get." You are never satisfied. Just look at you! You have enough clothes for another ten lifetimes! Why more? We all have more than enough things. If we packed them all together, we could not even carry them. We need a truck to carry all our possessions. Why more? Why do we think, if we only had the latest model of whatever, that would make us happy. When are we going to learn that happiness comes from giving, from generosity, from enjoying the happiness of others, and from contentment?

The terrifying propaganda that happiness depends on what we get is very dangerous. Not only is it destroying our planet, it is destroying our mind. Young people, little children, have all these advertisements on television. They all want designer clothes and designer toys. They are already plugged into this very insidious propaganda, which is increasingly violent and the opposite of any spiritual wisdom. When you watch the movies and games that children play, it is all violence! I read recently that by the time children reach the twelfth grade, they have watched an average of 32,000 murders, through their cartoons, through their movies, through the games they are playing. Every movie has to be more violent, more terrible, more gory, to get that extra little take.

All this greed and anger feeds this ego, this 'me'. I have to sell myself! I am the most important. If I am happy, then the rest of the world is okay. This ego, this adorning of the ego... these are the poisons, the poisons of the mind. No wonder, we are a sick society. Every day we are imbibing more and more of these poisons, then we are wondering why we don't feel well.

We have gone wrong somewhere. We have gone horribly wrong, and we need to get back to thinking about the basic essentials, and to our spiritual roots. Happiness rests in the happiness of others, giving happiness to others, not thinking so much always of our own satisfactions and benefits.

Our satisfactions and benefits are in giving joy to others, in being kind, in being generous, in being thoughtful, and in learning to cultivate our inner tranquillity, our inner clarity of mind, and our empathy with all beings. Not just human beings, but all beings. We can all do it. If others have done it, we can do it. But if we do not learn to do it, and if we do not teach our children to do it, if we give them all the wrong values right from when they are small, what can we expect for our next generations?

We are in dire straits. We can pull ourselves out, but we can only do this through transforming our own attitudes. Our attitudes are genuinely transformed through understanding. Understanding and loving compassion go hand in hand. The important thing is to transform our inner being,

The terrifying propaganda that happiness depends on what we get is very dangerous. Not only is it destroying our planet, it is destroying our mind.

because our inner state of mind is reflected in our outer reality. What is happening with our planet at this moment is a reflection of the beings inhabiting our planet, mostly the human beings. To transform the planet, we need to transform ourselves.

This speech was transcribed from an address to the Waldzell Conference in Austria in 2005.

Jetsumna Tenzin Palmo is a Tibetan Buddhist nun who lived for twelve years in a remote cave at 13,000 feet in the Himalayas, meditating twelve hours a day. She came down from the mountain and founded the Dongyu Gatsal nunnery in northern India to train Tibetan nuns in the spiritual practices that had been denied women for centuries, and to revive the female Togdenma lineage. Today her nunnery is thriving, and she has helped overturn the entrenched patriarchal barriers that keep women from rising to positions of spiritual authority in the Tibetan Buddhist tradition. She leads retreats around the world, and her inspiring life is recounted in *Cave in the Snow* (Vicki Mackenzie, Bloomsbury, 1999).

Llewellyn Vaughan-Lee describes the power of spiritual service in the outer world with its capacity to help others and transmute destructive energies in society. By awakening our consciousness we not only transform ourselves; we joyously merge with the world as its servant.

Spiritual Responsibility at a Time of Global Crisis

Llewellyn Vaughan-Lee

The physical body of the Earth has a spiritual energy structure, just as the human body has within it a spiritual energy structure. As is known by almost all spiritual traditions, the human being is the macrocosm of the whole. As we make this transition from a focus on individual spiritual practice into a global dimension of spiritual awareness – away from the individual back to Oneness – there is a whole dimension of spiritual esoteric science that will be unveiled. I say 'unveiled' at this time, because it has been known in the past in a slightly different way. You can see it, for example, in the European tradition of ley lines – spiritual lines in the Earth (known in the East as 'dragon lines'). Many important spiritual structures have been built at the intersections of ley lines, for example Stonehenge or Chartres Cathedral. There has been, in the spiritual body of knowledge that belongs to humanity, an understanding of the energy structure of the Earth, which is very real, just as your own spiritual energy structures are real.

As you know in your own spiritual development, certain energy centres are awoken. Unless these energy centres wake up, there is no spiritual evolution – it cannot happen. Part of the purpose of spiritual practise is to prepare yourself for the awakening of these spiritual centres, whether it is the heart chakra, the brow chakra, or finally for the full realisation of God, the crown chakra. The seven chakras are the best known of the energy centres of the body. But there are also other chakras, and at certain times in human history these energy centres have been revealed and then hidden. This has to do with the veiling and the unveiling of God, the continual process of the revelation of the Divine.

As you know in your own spiritual practice, these energy centres are not woken up by turning on a switch. They require a particular focus of spiritual energy that either happens through your own individual inner practise, or occasionally through the transmission of a teacher who has the authority and skill to open up an energy centre. For example, in our particular tradition of Naqshbandi Sufism, there is a way to spin the heart. There is a saying repeated by my sheikh from his sheikh's sheikh. When talking to a yogi, he said, "We are simple people, but we know how to turn the heart of a human being so it will take that person to where you cannot even imagine." This is done through a transmission of love that goes from the sheikh into the heart chakra. It activates the heart chakra, allowing for the development of a certain spiritual consciousness. This is all part of straight-forward spiritual esoteric science.

What is particularly interesting and very important is that the physical body of the world also has an energy structure. There are power centres in the Earth that need to be activated for humanity to evolve to the next level. Remember in your own spiritual practice that you cannot make a shift in consciousness unless a spiritual center is awakened. I have seen people do inner purifications for 15 or 20 years but nothing really happened because nobody awoke their spiritual centres. Remember the individual is a microcosm of a whole. The spiritual body of the Earth has these energy centres but they are not necessarily the same energy centres as before. I have been to Stonehenge and it is not alive any more – it belongs to a different era of human history. It is very beautiful but it is not awake. Even Chartres Cathedral is not fully awake anymore. It belongs to a different time.

There is an important spiritual work that belongs to the next era of humanity that cannot be done by any one individual or any one spiritual order. It is not allowed to happen that way. Why? Because the next era is about things coming together: it is about synthesis. It is not about one person, one spiritual order or tradition being separate from another. It is about how we can work together. Why is this so important? It is because the seeds of how a spiritual era comes into being will determine the character of that spiritual era. There is an important Naqshbandi Sufi saying, "The end is present at the beginning." When you begin something the ingredients then determine how something will evolve, what will come into being. There is important spiritual work in the world and it cannot be done by a single individual or just one group or tradition. It will not work; the doors of grace will not open. The energy will not flow. That is why we are here at this gathering: to constellate a certain energy through presence. Grace is given when people come together. By being together with this shared spiritual intention our spiritual awareness is aligned with the work. Remember that when there is a spiritual gathering, at least three quarters of what happens takes place in the unseen, on the inner planes. You know this in your own spiritual practice. Why do we go into meditation and close our eyes? So we can learn to be present and work in the inner worlds.

When we meet, we come together not only in the outer but in the inner worlds. We come together as bodies of light because that is what we really are. When we take spiritual work seriously, the Sufis say we polish the inner

...the next era is about things coming together: it is about synthesis. It is not about one person, one spiritual order or tradition being separate from another. It is about how we can work together.

mirror of the heart. Through the good fortune of my sheikh, I have been allowed to see what souls are like when they work on themselves, to see how vibrant and dynamic they are in the inner worlds. How full of light they are! When we remember God and when we are steeped in our remembrance, that light is amplified. A human being who gives themselves seriously to God and to spiritual practice is incredibly powerful. Why? Well, what is spiritual practice? It is a process of alignment. Mostly people scatter their energy throughout the day, in what is called in Taoism the 'ten thousand things'. People are caught in the myriad play of illusions. You know the Lord's Prayer, "Lead me not into temptation", yet He does just this with His wonderful world! We are led continually into temptation, into distraction, into diversion. Our energy is diverted and scattered. What do we do when we practise? We bring it back. We bring the parts back together. We take scattered awareness and make real focused awareness and our light shines much more brightly. Then we can be used by those spiritual beings who know how to use an awakened human being.

There is spiritual service in the outer world, helping those in need. This is an important part of spiritual service and loving kindness. There is also another dimension of spiritual service that hasn't been so widely understood: allowing your spiritual body of light to be used in the work for the world. When you are aligned with a spiritual tradition, whichever tradition it is, you are in the company of those who have gone before. The real *sangha* is not just us here, it is those who have helped us and also continue to help us from the inner worlds. In Sufism, there is a tradition of the *awiliya*, the friends of God who work for the spiritual well-being of the world. And also we believe that these spiritual Masters when they pass on are still present with us, helping us from the other side. These are the Masters of Love and Wisdom, our ancestors, and they are still with us. My teacher, Mrs Tweedie, used to tell me, "The Great Ones, the spiritual Masters, are extraordinarily powerful spiritual beings. They can see the past, they can be the future, but often they do not have a physical body. They need us, their willing servants, to be here for them. Then they can use us."

When you surrender to the teacher, the Path, it has very little to do with the teacher telling you what to do. This is a big misunderstanding that has caused a lot of problems in the West because we don't have a mystical context. What *really* happens when you surrender to the Path, when you surrender to your teacher is this: you allow your light, your spiritual body, to be used for his or her work (although there is no 'him' or 'her' at this dimension because this is the plane of the soul). There we can also work together. When we come together there is a constellation. There is an outer constellation – people travel to be part of the group and make a commitment to be there – and there is also an inner commitment, an inner constellation that creates a vehicle for grace. Christ said it very beautifully, "Where two or three are gathered together in my name, there am I in the midst of them" (Matt. 18:20).

There is a certain work that can only be done by different spiritual

There is spiritual service in the outer world, helping those in need... There is also another dimension of spiritual service that hasn't been so widely understood: allowing your spiritual body of light to be used in the work for the world.

Now we have to look beyond the individual and open ourselves to this wider dimension of spiritual science that includes an understanding that the Earth is a spiritual being...

traditions coming together, and part of that work is to activate energy centres within the spiritual body of the Earth. When we meet together at a conference, for example, there is a vortex of energy that builds up during the day. It is alive, it is pulsating, and it is singing the remembrance of God. We are not just here to listen to speakers, however fascinating or enlightening, we are here for spiritual work. We are here to be used by those who guide the destiny of the planet. They don't need a lot of people. Real spiritual work is very precise because it is done by Masters, by those who have guided the destiny of the planet for centuries. When you are part of a living spiritual tradition you are guided by someone who knows the workings of your own heart, knows how to give you more troubles than you think you can bear. The guide watches your reactions. Do you need love or difficulties? We are given what we need, the spiritual energy we need for our journey. If you only knew how we are taken care of – from the moment we turn our attention towards the Truth!... And those that look after us ask, how can this human being be used – not just in the outer world but in the inner world – because often it is not what I do, but how can I be? Doing is masculine, being belongs to the feminine. That's why my sheikh called Sufism a state of being. Yet we often forget to 'be'. When we are attuned, really present in the moment, in this extraordinary intersection between different levels of reality, between this physical world and the inner worlds, then we can be used. Then we can help to unlock these power centres in the Earth that are needed to change global consciousness.

This knowledge has always been part of the esoteric science of humanity, in the same way as our ancestors 500 years ago wouldn't consider building a sacred structure, a cathedral, unless they used sacred geometry. But this knowledge has been veiled from us. Recently, we have focussed on spiritual life as developing our own individual inner path, and we have gained some understanding of the spiritual body of the individual and its transformation. Now we have to look beyond the individual and open ourselves to this wider dimension of spiritual science that includes an understanding that the Earth is a spiritual being, pulsating with qualities of light. We have to develop a consciousness of the world as a whole, as a single living spiritual being.

When I began spiritual life thirty years ago there were hardly any spiritual books in the bookstore. There were one or two books on Buddhism and Sufism and that was about all. Now you try to find something in a bookstore, and you become overwhelmed. Why? Because the next step of humanity includes spiritual knowledge and consciousness. Over the last 500 years we have focused on developing our individual consciousness, and now we have to take it into the arena of spiritual work for the whole, bringing together individual and global consciousness. Our individual spiritual consciousness is a part of the spiritual consciousness of the whole, of life itself.

It is not enough anymore to be a solitary mystic living in isolation or living in an alternative society like some groups formed in the seventies. It doesn't work like that anymore. It has to be a commitment to life itself, to life as it is now: spiritual responsibility in a time of global crisis has to do with the moment. It is about becoming conscious that there is a need to respond to this moment. There is a need for us to participate as fully conscious spiritual human beings who are aware of the inner and the outer, of the in-breath and the out-breath, and to say 'yes'. To say, "Beloved, use me as only You know how to use me, because I am part of You."

We are all part of the living organism of life. We are not other than life. The moment the soul comes into incarnation, comes into a womb of a woman, it makes a sacrifice. The soul leaves behind part of its angelic nature to come into manifestation, into this drama of opposites which we know only too well because they cause us crisis and conflict. This is the drama of this world. There is a moment before incarnation when God asks the soul, "Am I not your Lord?" and the soul, looking towards its Lord, looking towards God, says, "Yes, I witness it." Then the soul makes a pledge to witness God in this world. It is a very powerful pledge because it entails bringing our consciousness fully into manifestation as a human being who witnesses God, who is alive for His sake. Remembrance of God is not just to remember 'I am remembering God', it is to incarnate one's divine nature which is remembrance of God. When you enter on a spiritual path you take on the responsibility to become aware of your divine nature, and then you have to incarnate it into all the mundaneness and conflicts of daily life. The practices of the path help you to do that. We can all sit in a cave, meditate and have wonderful pure thoughts but, as the lady said, once you get a speeding ticket, something changes.

At this present time we are allowed to consciously participate in the spiritual evolution of the body of the Earth. It needs us. The Earth needs our awakened presence to unlock its energy centres, to let the light flow. Just imagine what would happen if the light was allowed to flow in this particular city in America (Washington DC)... and I don't mean just the idea of people sending out good thoughts. You need a real power. This world has become so corrupted, so dominated by greed, power, and money, that a few good thoughts are not going to make much headway. But in the spiritual body of the Earth there are these centres of power, and again I go back to your own experience, to the model of the human being. You can do as much purification as you like, have as many positive thoughts as you like, but until your heart chakra is awakened not much is going to happen. That is why the Sufis place so much emphasis on the awakening or turning of the heart. Once the heart starts to turn it wakes up other energy centres. It speeds up your evolution. The light around you changes. You begin to attract different people to you. It is not generally understood, but you even attract different physical cells that spin faster; they have more light in them. The human being becomes more luminous. It then interacts with other human beings on the inner and the outer world quite differently. Things then begin to happen to you. You become part of the flow. Part of the Tao.

The same can happen to this Earth. It is waiting. That is why we are here, why we have gathered together at this time. Through us a certain energy can be given right into the spiritual body of the Earth that can unlock a certain door. These are the keys of spiritual awakening. They used to be understood locally only by shamans and other initiates. This is not something New Age I'm telling you. It is a very detailed, very particular spiritual science, and in the next era, much information will be given to us as we learn to work with the physical body in a spiritual way.

The coming together of spirit and matter is not some nice idea; it is very practical. Working spiritually with the physical plane has definite techniques. The way spirit impregnates matter, and how it changes the energy flow in matter, is a very particular esoteric science and this information will be given to us as we need it. Why? Because it is Oneness. The wonderful thing about Oneness is that it is one! Everything you need is here. People do not realise this because Oneness for most people is a concept, and they 'see' Oneness from the place of separation. Is there enough food for everybody in the world? It is one living organism! It is a multi-dimensional living organism just as human beings are multi-dimensional living organisms. The Sufis would say, "It is a revelation of God." It is a revelation of His Oneness, continually moving, unfolding, changing... that reveals Himself again and again and again, and He never reveals Himself in the same form twice. It is extraordinarily beautiful. Part of the difficulty of this present time of transition is that we still see through the eyes of the past. We still see it through the veils of duality so our attention is still taken where those veils guide us. We are drawn into the debris of a dying civilisation.

I don't know if you have had that experience, when suddenly the veils lift and you see a world around you that is in some ways the same world and in some ways totally different? I had my first mystical experience when I was sixteen. I was at boarding school and somebody lent me a book with Zen koans inside. I read one, "The wild geese do not intend to cast their reflection, the water has no mind to receive their image." That Zen koan was a key, it unlocked something inside me and the veils lifted. Suddenly this boarding school, where I felt isolated, lonely and had to study Latin, Greek and all sorts of things that had no meaning to me whatsoever, became suffused with joy and light, and I laughed and laughed and laughed. It was the most wonderful place I had ever been! It was a miracle, a continuing revelation. I would walk by the river and find it so beautiful, the light dancing on the water. For two weeks I saw the joke in everything, for in Zen practice everything is a joke. It is all an expression of the cosmic joke...

In Sufi groups there is also much laughter because the Beloved is always fooling us. He is always saying, you think it is like this and but once you identify with it, it becomes something else. He is the greatest trickster and he has created this extraordinary world of illusion that tricks us again and again and makes fun of us and tricks us some more. All this is but the veils of the Beloved. They seduce, deceive and betray us, but we can, through the grace of God, see it differently. Then you will see that what is around us is not the vestiges of a dying civilisation but a budding, an awakening of something incredibly beautiful.

The next era, when it comes into being, is unbelievably beautiful. If

Part of the difficulty of this present time of transition is that we still see through the eyes of the past. We still see it through the veils of duality so our attention is still taken where those veils guide us. We are drawn into the debris of a dying civilisation.

humanity knew what was waiting for it, it would throw away so much stuff. Do you need entertainment when you have joy? Do you need job satisfaction when you are really happy? It doesn't matter what you are doing because everything is an expression of divine Oneness. Everything is God's will. There is nothing other than Him. There is nothing other than That. It cannot be other, because He is not two. There is only one divine presence. This knowledge has already been given to humanity but sadly most people are looking the other way and reading their newspapers or watching reality TV. I just pray that we will be around when the next era really flourishes. It won't just be for our children or our children's children.

The other wonderful thing is that this shift in evolution is not a gradual shift. There is a very simple reason for that – we don't have time. If a global shift doesn't happen soon we will have ruined this planet. Any species that consciously destroys its own ecosystem does not have good prospects. There has to be a flip, a shift in consciousness, and we have to be the people that make the shift. Through our devotion, prayers, and spiritual practices we can bring into the world the light and love that is needed to help the world to transform.

In the depths of our heart we all carry a relationship with God that is love. "He loves them and they love Him." Practices such as remembrance, loving kindness, help us to bring this love into our relationships, into community, into the world. These threads of love then become woven into the fabric of life. What we don't realise, because we always see things through the physical eyes, is that these threads of love that come from our heart are like rays of light on the inner plane. They meet other threads of light and they weave themselves together into a fabric of light and love.

When a human being becomes spiritually awakened, when we become aware of our spiritual nature, we have a certain responsibility. Part of this responsibility is to be consciously aware of being a part of the living, breathing, spiritual dimension of life, part of this organic whole.

We think of ourselves as separate individuals doing our spiritual practice, trying to achieve something spiritually, because that is how we have been conditioned. But this is not our real nature. We are sparks of light in the fabric of love that is part of the world. That's why some teachings say that although we may think of ourselves as individuals, we are just the waves on the sea. We are all part of this ocean of life. We came here because the ocean of life and love drew us into the world, and we are responding to the need of life to become awakened, to become alive, to breathe again. Have you ever had that experience when suddenly something happens and you can breathe again, when joy starts to sing through you?

Imagine how this is for the Earth. Imagine if you were treated as just a physical object for a thousand years, and were continually raped and abused. What would this do to you? And then imagine if just one person recognised you for who you really are, for the beautiful being of light and love that is your real nature. On the spiritual path this is part of the role of a spiritual teacher. He or she recognises you for who you are, and something inside

We came here because the ocean of life and love drew us into the world, and we are responding to the need of life to become awakened, to become alive, to breathe again.

The song of the soul of the world will come alive in our hearts and sing to us and we will sing to it.

you starts to sing because you are recognised. Well, how would this feel for the Earth? Isn't it about time we recognised the Earth for what it is, not as a physical body with ecological problems, or even our Mother the Earth that we rape as much as possible. Yes, the Earth has physical problems but you know yourself that sometimes physical problems have a deeper cause because you are out of balance or not being nourished on the level of the soul. Something in you is not singing. There is no joy, no meaning, and nobody recognises you.

Once we step out of the paradigm of our own individual self, of our group, race, or culture, then we can step into a new paradigm, into a field of awakened consciousness, into Oneness. The Earth will respond because the Earth is waiting to be recognised – and not as some separate 'thing', not as some body that has physical problems, please! As mystics we know that the Earth is a revelation of God. It is not a problem to be solved. If you approach life as a problem to be solved, if you approach the Earth as a problem to be solved, if you approach this present time as a problem to be solved, you are going to get a problem. This is the law of intended consequences. But if we are present with an awareness of the Divine and we ask the Divine, "How do you want to use us in this unfolding of Oneness?," don't you think the Divine will show us? It wants to use us. This is so simple. We do not need complicated spiritual techniques. There is no 'problem'. The world has never been a problem. Often problems are created because we have to learn something. What is real learning? To make a shift in consciousness, a shift in awareness, a shift in a way of seeing, hopefully going into that state of real awareness where there is no duality, no subject and object, in which you can see things are they really are.

If you saw this world as it really is, you would just be amazed. You would be full of awe, full of wonder and you would immediately see what is your part to play, where your soul's destiny is meant to unfold because it is all written. This is the book of life, the book of love. It is all written here – how you can be of use – if you bothered to ask. If you are not so identified with your own problems, your own conflicts, and your own resolutions. What is true of the individual is true of the whole. When we see our world as it really is, our energy will be used for the sake of the whole. Our light will start to flow around the world and the physical body of the world will start to heal itself. I find it fascinating that we have reached an ecological crisis that is so bad we can't solve it on our own. But what happens with your own body when you become realigned? The energy starts to flow. Joy comes back. The body can heal itself and magic can come alive. This is true of the Earth. When it is realigned, the different spiritual centres all around the world will wake up and speak to each other. Then the world will begin to take its place within the solar system, in the galaxy, not as some isolated planet on the edges of the Milky Way that's making rather a mess of itself, but as a living spiritual being of which we are the guardians. The song of the soul of the world will come alive in our hearts and sing to us and we will sing to it. That is what we are being prepared for. It is incredibly beautiful.

I know what happens when the heart of an individual wakes up. I know the light that flows around it. I know the song that starts to sing in the heart because I have the grace to work with souls. I have been shown what happens. It is incredibly beautiful. It is a celebration. The angels start to sing and energy flows from the inner world to the outer world. This is the symbol of the New Age, the '8', the symbol of infinity, when there is a continual flow of energy from the inner to the outer and the outer to the inner. It is a flow of energy that will nourish and heal the world. It comes from the inner healing planes to the outer polluted planes and joy comes back. Then the energy centers in the world will wake up and they will throw off the dust of corruption and the debris of pollution. We can't do it individually but with the grace of God anything can happen.

We are here to help the world come alive with love. This is part of our duty as awakened human beings, as human beings who don't put their self interest first – either spiritual self interest or ordinary self interest, there is no difference really. What matters is service. That's why on the Sufi path after a stage of Union comes a stage of servanthood. We are slaves of the One and servants of the many. Because we do not want anything for ourselves – this is very esoteric what I am gong to say to you – we are allowed access to the places of power. These places are veiled and hidden from those who want something. Why? Because the power inside them is so tremendous, if somebody used it for their own personal gain it would be disastrous. If you think politicians have power, that is children's games. The power that is in the Earth and is waiting to be used is real power. It belongs to the will of God and the destiny of creation. It belongs to the whole purpose of evolution and life and there are certain things you cannot do without power, real power, power that belongs to the will of God. If God says, it will be, it will be.

The Sufis talk about the two poles – jamal and jalal. Jamal is the aspect of beauty, the feminine quality of God, and Sufis love beauty because we know how beautiful our Beloved is. Then there is jalal, the aspect of majesty, of power. Much spiritual power has been veiled, hidden, because it is so easy to misuse. This is not a power that plays out on the plane of action and reaction. It is not a power that belongs to duality. This is the power that belongs to Oneness. If something needs to be, it *is*! There is no process to this power, it just is. It functions. It makes things happen. This is the will of God, as is said in the Lord's prayer, 'May Thy will be done on Earth at it is in heaven."

How it is going to work in detail, I don't know. It has a lot to do with our level of participation and commitment. It has a lot to do with what happens when some of these energy centres are activated, awakened. Just as a human being can be given a mystical experience, and some are very disturbed by that experience whilst others are taken into states of bliss. If you withdraw back into the ego it is very dangerous. But if you give yourself in service to the whole, as a spiritual human being, then you can be aligned with the will of God and the work that needs to be done. It doesn't mean you have to start a project. I always advise people against leading a 'spiritual life'. It usually leads to disaster. Life is good enough as it is. There is no need to be 'spiritual' but there is a need to be a real human being. In fact, in Sufism there is an esoteric tradition of the qutb, the pole, who is the one complete human being in existence at any time, who lives the real potential of being a human being. The mystery is that we don't know that we are divine. Of course we are divine, we are made in the image of God. What else are we going to be? Just leaves floating on the wind, blown by the gales of consumerism? Please! All that matters is that we make a commitment, not to anybody else but to our Self, to our divine nature. There is a Sufi saying, "It is the consent that draws down the grace." When we *really* say yes, not a miserable sort of yes, but as a lover who says, "Yes, Beloved! I am here for You, I am Your lover, Your servant, Your slave. Do with me what You want. Show me how You want to use me. If You want to abandon me, abandon me. If You want to destroy me, destroy me. If You want to make me a CEO of a big company and drive a big car, do that!" Never limit the ways of the Beloved. Never limit the ways He can make a fool out of you, or turn your world upside down. Whose world is it? Does it belong to the lords of worldly power, or does it belong to the Lord of real power?

When the world starts to sing, when the heart of the world starts to open and the world starts to sing again, it is going to be so beautiful. I know what happens when a human being's heart opens and when a human being's heart starts to sing. The angels turn their attention and look at that human being, because they hear the singing of the heart. They hear the heart praising God and the angels bow down. When a human being awakens and the heart sings the praises of God, there is nothing more beautiful in all the worlds. If that happens to one human being, think what will happen when the heart of the world awakens and starts to sing.

Transcribed from a talk given at the 2003 Conference on Spiritual Responsibility at a Time of Global Crisis, National Cathedral School, Washington, DC.

Llewellyn Vaughan-Lee is a Sufi teacher in the Naqshbandiyya-Mujaddidiyya order, and the successor to the Sufi master Irina Tweedie. He is the founder of the Golden Sufi Center in Pt. Reyes, California, and is the author of over a dozen books including *Working with Oneness, Alchemy of Light,* and *Prayer of the Heart in Christian and Sufi Mysticism.*

MARTI describes how a tremendous explosion in consciousness is occurring on our planet at this time, thrusting us into new dimensions. She describes how a new collective consciousness is manifesting in intentional communities, and in other places throughout the world.

Living in Auroville: A Laboratory of Evolution

MARTI

A tremendous explosion in consciousness is occurring on our planet. This explosion is thrusting us into new dimensions. We are being carried on a lucid thread of awakening to a potential that is new, yet as ancient as the stars. This new collective consciousness is manifesting in intentional communities, and in other places throughout the world.

Our Planetary Mission

We didn't just arrive here by chance. Each one of us has a mission on planet Earth, a mission that can carry our spirit to its true destiny by resonating at higher and higher frequencies until it realizes its true potential. In human terms, this is the energy field that recognizes everything as one, that sees all events and beings as equal and lives in the light and love of compassion. It is the energy that connects to the deepest recesses of the Earth, and the farthest realms of the stars. It is the tempest and the still lake. It is the consciousness of everything and nothing. It is what isn't and what is. It is the seed out of which the plant is born and the plant that gives birth to the seed. It is the healer and the healed. It is the pure joy of existence, the conveyer of unity, the evolution of evolution, the gateway to *point bindu*, that miniscule dot in the universe from which we can see the entire universe. Every sentient being on Earth has come here for a purpose. Nothing exists in a vacuum. Every force field in the universe is interrelated. And our mission on this earth is to manifest this consciousness in our lives.

One of our challenges as human beings is to connect with our *legende personnelle*, or personal destiny, as Brazilian writer Paolo Coelho calls it in *The Alchemist*. It is a song line that will lead us on our quest for the inner truth of being, our true *raison d'etre,* and we will become part of an energy field of great significance. This collective consciousness, described by great mystics of many traditions, has an exponential power to change the energy patterns of the entire universe, like a stone thrown into a pool that ripples and expands forever.

The Cosmic Shift

The Indian philosopher Sri Aurobindo has called humans 'transitory beings' and describes a time in the future when a few highly evolved beings will experience a mutation of the species, or a deep shift in collective consciousness. Through this giant evolutionary leap, humans will truly begin to experience a 'divine life force'. The Russian philosopher Gurdieff referred to this collective potential when he said that if one hundred conscious people were meditating together on the same wave length our world would be a different place. Our Mayan ancestors say that in the year 2012, a great human cycle that began thousands of years ago, will come to an end, and a new consciousness will be born. How this will happen will be clear in time, but what is certain is that today evolutionary changes are happening at a highly accelerated pace. As the writer Peter Russell reminds us in the *Global Brain* and *The White Hole in Time*, consciousness is resonating at faster and faster and more subtle frequency levels. In effect, more significant changes have happened in the last 30 years than in the previous 500 years.

Return to the Mountain

Living in communities that have a strong connection to the Earth, we are reminded that nature is our greatest teacher. Nature connects us to our deepest sense of who we are. In many cultures, the mountain is the sacred abode of the divine. Climbing the mountain is visiting our highest dwelling place, where we connect to our deepest aspiration, whatever our spiritual tradition. As we listen to the heartbeat of Mother Earth, we find our rainbow wings. It is a place that connects us to the Earth's energies, where we can respect and love ourselves, see the horizon, and give up our attachment to those things that weigh us down and keep us from realizing our true spirit. The Chinese say, 'Embrace tiger, return to the mountain'.

There is a beautiful Navajo (Yebechi) chant: *IN BEAUTY, I WALK… All around me my land is beauty, IN BEAUTY, I WALK.* As we sit on the summit of a high mountain where the wind blows fiercely and we feel the sky penetrating our bodies, we forget the fight we had over our neighbours' fence. We start to grasp the true power of existence. When the empty sky becomes as full and dense as the forest of trees below it, we begin to feel space as moving, ethereal, generous, holding all possibilities. We understand that no object has a value beyond our perception of it. We are pilgrims. We know that the Earth is our home, but so is the entire universe. We are wisdom-keepers. We are ageless. We live in the age of agelessness. This is the deep innate wordless knowledge of every spiritual tradition. When we walk in sacred landscapes, that awareness is awakened. We are empowered to co-create our reality, to realize our true destiny. The Buddhist monk Thich Nhat Hanh suggests that each conscious step in walking meditation leads us closer to true awareness. Establishing sanctuaries, where we can hear our own heartbeats, where we can rejoice in being together and listening to nature's song are at the core of our well-being. Restoring the Earth becomes one of our primary goals.

Each of us has been given a soul, a body, and a mind that serve as a vehicle. In India, we say, 'Nada Brahma', or 'sound is god'. The Indian mantra, the Gayatri, or Ode to the Sun, which is sung each morning and evening by millions of people, is a prayer for illumination. The sounds of this mantra vibrate to work on the different organs of our body and to transform our minds into a wonderful state of well-being, what Gurdjieff would call the 'Act of Self-Remembering'. What better way to practice Nada Brahma than in nature, to understand it in the spin of our tiny Earth in space, in the OM intrinsic in our breath as we take the universe inside our bodies and breathe it out again?

We Are One Body

We are all unique, yet we are all part of a web of life that is interdependent; a rushing stream hastens towards the ocean that refuses no river. When we acknowledge that we live in community, we dissolve into One Body and not only become part of Mother Nature who has nourished us and gives us life, but part of a universal cosmic consciousness that contains everything that we have been, are and will be. This consciousness is beyond time and space. It cuts through the vast and the minute, the within and the without, the one and the many. It is akin to dolphin mind. It is lucid, flowing, ephemeral as waves, swift as a falling star, magical as the smile of a newborn baby. It is that inexhaustible energy that is generous because it knows it is abundance itself.

We are all unique, yet we are all part of a web of life that is interdependent; a rushing stream hastens towards the ocean that refuses no river.

Great Philosophies All Merge

Mahatma Gandhi once said that all religions are like leaves on the same tree. Unity is at the forefront of all timeless teachings. For thousands of years ancient Taoist, Buddhist, Christian, Jewish, Moslem, and Vedic wisdom has emphasized our Oneness. Zen Buddhism shows us that separateness is an illusion. The Upanishads state that if any one being suffers, that energy carries to all beings. His Holiness the Dalai Lama says, "The basic fact is that all sentient beings, particularly human beings, want happiness and do not want pain and suffering. On those grounds, we have every right to be happy and to use different methods or means to overcome suffering and to achieve happier lives."

Christianity suggests that we are all brothers and sisters of the same body and blood. The Koran states that "All things in creation – whether concrete or abstract – are but shadows depending on His Light." Western science, be it relativity theory, holographic science, morphic resonance, chaos, or quantum theory, emphasizes that we cannot function in one part of the universe without affecting entire fields. In physics, the dream of finding a unified field theory, which is symbolized by the research of astrophysicist Stephen Hawking and other cosmologists, is part of trying to understand our full dimension.

The Matrix

In today's world we confront the pressure of the Matrix, a programme or set of connections or synapses, ruled by the mind. The purpose of the Matrix is to push us to make a choice. One alternative is to choose the soul and the light and love of compassion and peace, or what the Bhagavad Gita describes as our 'Krishna', or 'Christ energy', that is who we really are. The other alternative is to succumb to the desolation of the heartless mind, or the force of Karna, and karma, or what others want us to be. Both forces live in the energy field that presently surrounds Planet Earth. Both forces are within each and every one of us. Ironically, without the Matrix, we cannot manifest the depths of our true consciousness, because it is an energy field inside us, as well, and we are all part of everything that exists. Yet it is lucidity, not fear that allows us to remain free. Nelson Mandela once famously quoted Marianne Williamson:

> *Our deepest fear is not that we are inadequate. Our deepest fear is that we are powerful beyond measure. It is our light, not our darkness that most frightens us. We ask ourselves, Who am I to be brilliant, gorgeous, talented, fabulous? Actually, who are you not to be? You are a child of God. Your playing small does not serve the world. There is nothing enlightened about shrinking so that other people won't feel insecure around you. We are all meant to shine, as children do. We were born to make manifest the glory of God that is within us. It's not just in some of us; it's in everyone. And as we let our own light shine, we unconsciously give other people permission to do the same. As we are liberated from our own fear, our presence automatically liberates others.*

As they say in Africa (Yoruba), "Eyes that have beheld the ocean can no longer be afraid of the lagoon."

Unity Through Diversity

As James Lovelock points out, variety and differences in nature are not just tolerated. They are actively encouraged and supported. Biologically diverse systems are the most stable forms of life anywhere. Obviously our very human survival depends on Mother Earth and her planetary diversity. And so we nurture the paradox that we are uniquely different from every other living being on Earth and at the same time intrinsically interconnected. The rich biological nature of our existence is strengthened when we live in community. When we live in mutual interdependence, we see that what one person cannot do, another can. This empowers us. We become many times more effective in a group than individual energy or capacity would permit.

Exercising Memory

Memory is the capacity to remember who and where we have been and to bring a past experience into the present. It is an energy field that helps us to situate ourselves in time and space and record the quality of our experiences. Tibetan Buddhists are masters of the art of harnessing memory to evoke consciousness. Through dream work, they train their minds to remember valuable information because we are what we think. Lamas also train to remember complex texts said to have been studied for many years in former lifetimes. This is why meditation, collectively working on our *sadanas*, or spiritual work, observing our thoughts, and cleansing our auras, is important. It is one way our memory serves us.

Japanese researcher, Emoto Masaru, has studied the molecular structure of water when it experiences different states of meditation, or discordance. When water is pure and experiences harmony, its molecular structure registers great geometrical beauty. When water experiences dissonant noise or pollution, its molecular structure becomes fragmented and ugly. In *A New Science of Life*, biologist Rupert Sheldrake demonstrates that in morphic resonance, when a critical mass of consciousness is reached in a specific species, the collective memory of the entire species is activated.

Some of our experiences with memory reinforce old fears and old lives because they are purely mental forms. Real freedom comes when we step outside the prisons of the past. It comes when we let go of the structures and material symbols that overpower our true sense of identity and keep us from realizing our true potential. Each of us has been born into a certain culture, race or family that we will probably never shed. But seeing this identity as an envelope, a vehicle that facilitates our rights of passage, helps us to overcome certain obstacles. It gives our spirit the freedom to awaken to the reality that we are all one body and that we are here for a real transformation beyond how and where we were born.

Real freedom comes when we step outside the prisons of the past. It comes when we let go of the structures and material symbols that overpower our true sense of identity

The Way We Learn

Living in communities gives us a unique opportunity and challenge to listen receptively to each other. Communities in natural settings help us to learn how to listen for the power of sound, how to recognize and understand natural patterns, how to read the landscape and skies, and to listen to the land. This means taking care that our human relationships are in harmony and co-creating an ecological footprint on the Earth that is light. When we allow our minds to absorb our ecosystems and to vibrate with them, then our cells become like living water. They are clear, transparent, and highly resonant. The Mother, Sri Aurobindo's spiritual companion, describes this process in detail in the *Mind of the Cells*. If we truly understand that we are one body, then we know that we cannot harm each other without harming ourselves. Love and caring become our sole real purpose for existence. Because we love ourselves and care deeply for our planet, simple living patterns become our way of life. Thich Nhat Hanh says, "If you are in a good community, one in which people are happy and living deeply each moment of their day, personal transformation will take place naturally, without effort..." This is highlighted in community where we play together as children and later see our own children and then grandchildren playing together. When we live in conscious communities our children no longer '*belong*' to us, they develop more easily as people of their own right and they learn about true generosity of spirit.

> *Community living can challenge are deepest beliefs about ourselves and shatter our egos. It provides a wonderful opportunity to see obstacles, for whatever reasons, as priceless lessons...*

Community living can challenge are deepest beliefs about ourselves and shatter our egos. It provides a wonderful opportunity to see obstacles, for whatever reasons, as priceless lessons. If we see everything, including our experiences, only as tools, or instruments for our evolution, and we learn to be detached from them, then we become enlightened beings. We feel a deep sense of identity with all that lives and a deep sense of happiness and gratitude in being part of something indescribably wondrous. This is the joy of compassion (*com+passion*). This is the Krisha, Christ, or Maitreya Buddha energy that simply radiates pure love because it is conscious of its own existence. It is beyond differences and the competitive values that enslave our egos. Our spirits yearn to live in love, peace, and harmony in every aspect of our existence. As we share this energy with others, the *com*, or coming together of *passion*, or intense love for life and existence grows. We experience the world in truth, honesty, generosity, equality, and deep compassion. We will radiate a strong experience of all being part of One Body. And this resonance pattern will reach out into the universe and carry us into new worlds.

Homing in on Heyoka

Vision quests and the medicine wheel are integral tools on the journey towards evolution in the Amerindian tradition. In *Buffalo Woman Comes Singing*, Brooke Medicine Eagle describes the pathwalker as a rainbow child. The journey begins in the south where the seeds of life are planted and walks to the west where culture and experience are learned, then turns to the north where knowledge and experience become wisdom, then rotates to the east where illumination occurs. After all four geographical directions have been explored, the initiate steps into the centre to experience the Earth and cosmos as a whole and the realm of within, or the *heyoka*. In the Lakota, or Sioux tradition, the ultimate consciousness manifests as the clown, or awakened spirit who laughs at everything because nothing is sacrosanct and everything is relative. The *heyoka* walks behind us and mimics our every action to expose our seriousness for what it really is: a cage we create for ourselves so that we are not really free to resonate on a deeper level. Being able to make fun of everything from the inside helps us to break up our fear and see it for what it really is.

Nothing is so serious that we cannot move beyond its absurdity to see the paradox of our existence. In this life, we have to confront and master all states. It is part of our evolutionary process. Love and harmony are words that describe a connection. Yet our experience of love exists not only in our relationship to ourselves, but to others. Paradoxically love does not exist without the notion of an absence of love. Our perception of harmony does not exist without a perception of discordance, as well. The sweet melody of a flute is sweeter when it makes its way through distinct patterns of chaos that shift and change and then become their opposite. Understanding this paradox gives us the distance we need to experience the laughter that comes out of cosmic awareness: the simple but profound understanding that despite all our differences, we are all really one. As the ancient Zen proverbs says, "That which is, is and that which isn't, also is."

Awareness of Awareness

Individuals living in community and sharing the same belief patterns, form a collective consciousness that can define or reshape the world. We create possibilities by believing that they actually do exist and we dissolve limitations by experiencing ourselves as being without limitations. Nothing is impossible. This is because, in actual fact, the universe arises within awareness. It is not that we dream a universe. It is that we dream the universe into existence with our very awareness. There is no inside and outside. There is simply an awareness of what is. Our source of life is neither separate nor contained. It simply is. Consciousness does not evolve from the universe. The universe evolves from consciousness. In the West, we tend to localize the soul in the body. In the East, the body is part of the soul and is seen as an instrument for transformation and change. Indeed, our bodies are the instruments through which we change our mindset, and transform our consciousness down to the very depths of our being. Communities are a way to practice that consciousness and 'bring it into matter', as Sri Aurobindo would say.

Listen and Trust

The experience of being alive, of giving, is our greatest gift to the universe. The *tat,* the one existence, or *tsd ekam*, in Vedic terms is our joy, our *raison d'etre*. Desiring and resisting require effort. Accepting and appreciating are effortless. With the stillness of the mind and body that comes through inner discipline, meditation, yoga, or simply by breathing in awareness, we extend trust to others. An inner transformation leads to outer possibilities. Everyone dies one day. The only real difference between any of us at that moment may be our level of consciousness. With a touch of humor, Japanese haiku master Kaiga wrote in his death poem, "Strange – like messengers they fly left, fly right, the fireflies."

Let us honor all that we are. Let us honor our love for beauty, our need for peace, our yearning for freedom, our belief in the existence of higher forces. Let us respect the necessity for compassion, our search for balance and meaning, our fierce need to develop courage and true integrity. Let us honor all of this to envision the future, and to set the creative forces in motion. We will take risks and made mistakes, but it will be a great adventure. As Lebanese philosopher Kahlil Gibran has said, "That which is boundless in you abides in the mansion of the sky, whose door is the morning mist, and whose windows are the songs and silence of night."

We are pathfinders but we are also dreamers and dreams seed wonderful things. They seed a certain wildness of mind, as we emerge and become participants in the unfolding of our own potential. A certain beauty appears in the web of bonds between humans, the awareness of awareness, when we understand that ideals are just tools, that delight and dismay have the same roots, that we are connected. That we have a third eye, a third ear. That our bodies are our temples of wakefulness. That inner stillness reigns.

We Are Evolutionaries

We don't start out to be pathfinders. We live our lives to be ourselves. The expression of our integrality is what leads us to become evolutionaries. As the early ancestors on the rugged shores of New Zealand have said, "I am, Io Mata Ngaro, the Unseen, the Creator of All, the Keeper of the Silent Space, the Singer of the Song that holds the Stars in Place and All that has Ever Been and Ever Will Be." When we have truly become ourselves and manifest a passion of a deep vision from a genuine place, then we become pathfinders. As the French writer Antoine de Saint Exupery once said, "…it is only with the heart that one can see rightly. What is essential is invisible to the eye."

What an exciting time to be alive. We are becoming the change we seek in the world.

MARTI from Auroville, South India has 30 years of experience in the field of education, 20 years of community experience, and is a former professor at the Sorbonne University in Paris. She is co-founder of 'Children and Trees Research' with UNESCO and the Indian Government. She has also worked on the Earth Restoration Corps curriculum and The University of the Streets and Alleys. Her current concerns: establishing a wildlife sanctuary to protect a UNESCO World Heritage site in central India.

Hildur Jackson is one of the early founders of the ecovillage movement, and before that the co-housing movement. This piece articulates not only Hildur's perspective on the emerging worldview, but is also an autobiographical account of her generative role in the ecovillage movement.

Who Am I? Why Are We Here? Living the New Worldview

Hildur Jackson

There are two states of being: one with a closed heart and one with an open heart. Love is the opening of the heart. The worst truth about the patriarchal era is that it prevented love. The misery of western society lies in the fact that no permanent love is possible, because from very early on open hearts are bombarded with unimaginable disappointments and meanness. At first children have open hearts, but little by little, they close them off because adults today usually have no idea how to deal with open hearts… A humane world can only come about through open hearts.

Dieter Duhm, Tamera, 2012

A Personal Quest over 50 Years

Like so many others I started wondering early on in life: who am I? This foundational question propelled me on a personal journey that I recount below, in hopes that it will shed some light on the major challenges of living in our Danish society. I hope also that this personal account might help people in other parts of the world to understand the limitations of Western culture. We all have to go through a similar process to start living a new worldview. This is now brilliantly described by Spiral Dynamics.

It all began with my role as a woman and later as a mother. My own mother often complained that the human condition was merely 'the lust of the flesh and the irreparable loneliness of the soul' quoting a Swedish writer Hjalmar Soederberg. My instinct knew it was wrong but her despair crept into me as I went up the hill behind the house at night and watched the stars. I had no words to help her. Her suffering was caused by the old worldview

Poster for the Nordic Alternative Campaign

and the suppression of the feminine for millennia. As a suburban housewife, she felt dependent and lonely, like most of the marginalized and frustrated housewives at that time. This was just after World War II. I was afraid of falling into the same trap, so I promised myself at the age of 14 never to marry and to get a good education. Life became a voyage to find a different understanding (worldview) and a different lifestyle.

Yet I still needed to find out why I felt totally alienated from mainstream Denmark – not wanting a professional career nor wanting to continue the marginalized role of my mother's generation, leaving me with two equally poor choices. I wanted to create a better and more loving relationship with my husband than the examples I saw around me. I was lucky to get an open-minded husband and find myself in the midst of many others having the same ideas.

After completing law-school, I started studying social anthropology in 1968 having just had my first child. I co-founded the 'redstockings' (the name of the Danish women's liberation movement) in 1970 and later worked with the 'ecofeminists' and 'The Flying Women', which were something in between witches and angels, defining that women had something important of their own to contribute. Together with other women I came to trust my own experiences, a deep belief in love as our birthright and a search for a new culture.

Cohousing and ecovillages was one answer (which I still fervently believe in), which arose as a women's project but also as a global and ecological solution. We lived in a cohousing for 20 years and co-founded two ecovillages.

Another project was the 'Nordic Alternative Campaign', which united 100 Nordic grass roots organisations with the scientific community to create a global vision where the global, the social and the ecological problems were solved. It led us to redefining our worldview and also seeing patriarchy as a major problem. Mats Friberg came up with a new paradigm in social research: VETA (science needs Visions, Empirical Studies, Theories and Action) and edited three books with Johan Galtung: *The Crises*, *The Social Movements* and *The Alternatives*.

Parallel to these outer endeavours was a sustained spiritual quest over more than 30 years. Reliving my own birth during 1-2 hours from listening to the Canadian Harmonic Choir, and doing holotropic breathwork with Stanislav Grof were revelatory experiences that satisfied my need for personal proof of the divine presence. A decade long friendship with William Keepin, and more recently learning from my eldest son Rolf, has added to this and given me several personal experiences verifying that a holistic, spiritual worldview is the correct one. The materialistic worldview has had to yield.

A Danish spiritual teacher, Jes Bertelsen, was the person who introduced me to meditation, chakras and dream interpretation. I am grateful for his work bringing a big group of people in Denmark to a new level.

My spiritual search was, however, always influenced by an experience (which I have only heard of a few other women having) of actually being the whole Earth. It happened at Eastertime during a meditation and is always with me. This can be seen as a continuation of the feminist approach and may be why tantra seemed natural to me and has acted as the motor of my spiritual development. Practicing tantra with my husband created a deep bonding which ordinary sex probably would not have done. Living, raising a family and working with him for 45 years has been a lot of fun, a constant learning process and a real blessing.† (Footnote: Jes Bertelsen and his former tantra partner have written substantially about tantra in Danish. For literature in English see Stephen Wik's book, *Beyond Tantra* or works by Georg Feurstein or Daniel Odier)

Another powerful meditative experience introduced me to the Christian mystic, Giordano Bruno. I have been studying him ever since. He showed me how and why religion and science parted when the Catholic Church burnt him at the stake in 1600 for his belief in reincarnation and his wish to create a religion of Love.

For me, Rashmi Mayur, an Indian friend whom we worked closely with for many years till his early death, was a reincarnation of the spirit of Bruno, having this time the whole world as his arena. He advised the UN and many governments of the South. He was opposed to religion but was nevertheless a very spiritual person. And a devoted scientist. My yearlong daily contact with him taught me a lot about the South and about spirituality. When meditating, a poem would often come to him in finished form.

The Christian mystic from Cyprus, Stylianos Atteshlis, also known as Daskalos together with the Danish spiritual teacher Asger Lorentsen, confirmed the reality and validity of my Christian roots. The American theologian Thomas Berry gave this a new philosophical dimension with his Dream of the Earth, a celebratory worldview, when we spent a week with him in Assisi. Thomas Keating, a US Catholic priest and teacher, is reinforcing this.

Another important influence has been the spiritual Buddhist culture of Ladakh, which I learnt about from Helena Norberg-Hodge when she was writing her book *Ancient Futures*; about Buddhist philosophy from Ari and Vinya Ariyaratne (Sarvodaya) and from SEM, Spirit in Education Network based in Wongsanit Ashram in Thailand.

In 1992 I met Chari, the current head of Sahaj Marg, a Raja Yoga system rediscovered in the 20th century. I have been a member of this meditation system ever since. Chari lives a new worldview and teaches how to open your heart to a growing group of people. Sahaj Marg has spread to 90 countries, and provides a practice with all the necessary elements for transformation of consciousness and a social context for spirituality (see www.SCRM.org).

Mentioning all these traditions such as the female/tantric/indigenous; Buddhist; Christian; Hindu/Sufi Inspired and a more science based modern approach illustrates that they all contribute and are relevant paths. They do not contradict each other.

A Synthesis

Eckhart Tolle is a Western equivalent of an Eastern enlightened master, expressing eternal truths in a language close to science and more easily understood by a Western audience than many Eastern teachers. His teachings have since spring 2008 become accessible all over the planet for free as YouTube videos, www.Opraheckhart.org

I use the teachings of Sahaj Marg and Tolle's recent book, *A New Earth*, as the bases of the following:

We are spiritual beings with access to divinity. To the question: Who are we? I have come to the view that we all have the divine light in us. It is our birthright. It is our true nature. We are all one. I now like the term: A Worldview of Oneness. This is the essence of the new worldview and this is what spiritual systems of many orientations tell us. The purpose of life is to express love in all its manifestations. Love is the expression of unity in a diverse world. By transforming our consciousness, we can become co-creative with evolution itself. Liberation and unity with the whole are further steps. Buddhists might call it reaching total emptiness or nirvana. We are reflections of the macrocosmos and have access to all the information and knowledge we need. The perennial philosophy is rich and diverse and enlightened Masters down through the ages have shown us the possibilities and the path to follow.

In virtually all spiritual paths, one of the essential tasks is to clear away whatever obscures or dims the divine light in us, and to remove the blockages in our minds and bodies created in this lifetime or in former lifetimes. This cleansing is done in different ways in different traditions, either on one's own or with support from teachers or the Master of the tradition (whether living or not). Regardless of the different terminology used in various traditions the cleansing process (e.g. samskaras in Hinduism, or purification in Christian terms) afterward the natural essence of the inner self shines through unabated. Eckhart Tolle defines what obscures us as a pain-body, which is not really us (being spiritual beings), but which can take us over very easily. The pain-body has been created by emotions of pain, anger, fear.

Everything is Vibration

The whole human system functions automatically and links us to the universe. We breathe, blood circulates, we grow and renew our cells without conscious effort. Even when the body has been functioning 69 years (my present age) it is capable of getting back to the original design if you let go

The purpose of life is to express love in all its manifestations. Love is the expression of unity in a diverse world. By transforming our consciousness, we can become co-creative with evolution itself.

of blockages ('samscaras' in Sahaj Marg, or the painbody in Eckhart Tolle's language). "All things are vibrating energy fields in ceaseless motion", says Eckhart Tolle. Matter is an illusion. Our senses perceive vibrations as solid and motionless. We are made up of molecules, atoms, electrons and subatomic particles, which are vibrating together, creating what we perceive as solid.

Once, while sitting in meditation in a forest with a dear friend, this became reality to me. I could see other faces around his face as clear as the natural one, like holographic images. But I could only focus on one at a time. Because of the meditative state I could also see clearly through him to the bark of the tree behind him.

Energy is vibrating at different frequencies. Thoughts consist of the same energy but vibrating at a higher frequency and with less density than matter, which is why they cannot be seen or touched normally. But that is not quite true – I have seen Rashmi's poems as vibrating colours – and I was never mistaken when I asked whether a poem had come to him. Thoughts have their own range of frequencies with negative thoughts at the lower end of the scale and positive thoughts at the upper end. Thoughts are very rapid vibrations. "Matter vibrates very slowly. Thoughts affect the slower vibrating systems like emotions," Will Keepin says, "Love creates form. If you resonate with the highest vibration you let spirit lead." Spiritual and natural laws are then one.

"Your body is inseparable from universal intelligence as one of its countless manifestations," Eckhart Tolle explains. "The atoms that together make up the body give it a sense of cohesion. There is an organizing principle behind the organs of the body, the conversion of oxygen and food into energy, the heartbeat and circulation of the blood, the immune system that protects the body from invasion, and the translation of sensory input into nerve impulses that are sent to the brain, decoded there and reassembled into a coherent inner picture of outer reality. We do not run our bodies. We are not in charge. The divine, God or the universal intelligence does this. It is the same intelligence manifesting as Gaia, the complex living being that is planet Earth." This is true for any life form. It is the same intelligence creating all circulatory systems both in our bodies and in nature. The chakra system is one way to define this. You may distinguish between the physical and other circulatory systems in nature and in the body as expression of microcosmos and macrocosmos.

Healing our Thought Patterns

Thoughts are always from a lower level than spirit. Thoughts need to be expressed in language. It is the nature of language to separate in good and bad, black and white, false and true and it has rules for how to combine words. Words, sentences and language are dualistic in nature. For that reason they can never be truth. Language should venerate nature; it is our task to create and use such language. Poetry is close to intuition and can express what has not yet been fully thought. Through poetry we may touch spirit.

Thus thoughts can affect and change matter. This is rather difficult for our materialistic society to understand but Raja Yoga and other forms of mysticism have known this for thousands of years. The Raja Yoga teachings are: We all have the divine spark in us and access to it. Impressions, thoughts from this and earlier lifetimes have dimmed this light, but we can remove them. We think with the whole body with all our impressions and blockages.

Many of us repeat the same thoughts day after day, year after year. The implication is: we must all be responsible for our thinking and for removing the old impressions. According to an old Danish saying, "Thoughts are toll free." But that is clearly not true. Every thought has consequences. It remains, and does not disappear. Just the thought alone can create some negative results. We must learn to think in a positive holistic way if we want the Earth to become a better place. We must develop our consciousness. Sahaj Marg has developed a cleaning technique to do this. A person cleans herself every evening by using willpower to imagine grey smoke leaving through the spine and then to imagine divine light filling the vacuum. And she goes to a preceptor and gets cleaning and transmission from the spiritual Master. A prayer and constant remembrance is a way to constantly stay in touch with spirit and be present in the now. Having a living spiritual Master is a big help.

Healing the Pain-body

Tolle teaches that emotions are the body's reactions to thought. He defines the 'painbody' as a conglomeration of challenging emotions created in life through fear, anger, jealousy and habitual behaviours. Other spiritual systems have different terminology to describe the same phenomenon (samskaras and vasanas in Indian traditions, nafs in Sufism). The painbody functions like a negative inner personality, a kind of second nature, which is taking us over at times and reinforces wrong thinking and destructive behaviour.

The pain-body, however is not just individual in nature. There is also the 'collective painbody', comprised of the shared pain suffered by countless humans throughout the history of humanity. It is fuelled by example by a history of continuous tribal warfare, of enslavement, rape, torture and myriad forms of violence. This pain still lives in the collective psyche of humanity and is being added to on a daily basis. Women suffer a collective female pain-body as do tribes, nations and races. The suppression of the feminine principle especially over the past two thousands years has enabled the ego to gain absolute supremacy in the collective human psyche. Although women have egos, of course, the ego can take root and grow more easily in the male form than in the female. This is because women are less mind-identified than men. We have to weaken and take energy away from the ego and make it powerless.

The first step in overcoming the painbody is to accept that it is not the true Self. The next step is to stop feeding it, so that the painbody withers and eventually dies. This entails developing an awareness to recognize when the painbody takes over and to stay present while completely dis-identifying with it. Tolle sees it as an entity not being the higher Self but closely connected to the ego.

In transmuting the painbody and removing false thoughts and outworn impressions, the energy centers of consciousness in our body, mind and heart are cleansed and opened. This in turn enables us to release the past and focus on the heart, so we become centered and inspired once again in our thinking and action. This system of subtle energy centers is known by different names in different traditions such as chakras in Indian philosophy or the Tree of Life in the Jewish Kaballah.

The Chakra System

In describing the person from an energetic point of view, the chakra system is a useful supplement to the above. It gives us a way to describe how thoughts are energy entering the whole body and affecting it. The purpose of the Danish Ecovillage Network in 1993 was defined as follows: To restore the circulatory systems on all levels in people and nature. All levels referred to the seven levels of the chakra system. In this way, the inner and the outer were united. Something similar was defined by Ken Wilber as 'integral practice' in his book *A Theory of Everything*. Thoughts create matter. We think with the whole body. Not only with the mind. I have developed a clear feeling of where my thoughts come from – whether they come from the heart or from the stomach (solar plexus). I have for years realized that I cannot trust the thoughts coming from the stomach as being from my higher self. They are always wrong and bring bad effects and decisions, i.e. painbody emotions. I have found that it is best for me to wait with important decisions until I have 'cleaned' myself (in the Sahaj Marg sense). But when I am centred in my heart, I know I can trust my thoughts.

The existence of the chakra system is widely recognized in many spiritual traditions, but what is far less widely understood is the Earth itself also has a 'chakra' system of subtle energy centers. Due to the holistic unity of life, the human body is a microcosm of the macrocosm of Earth. The vital consequences of this are addressed by Llewellyn Vaughan-Lee (see pages 63-72).

Thoughts create matter. We think with the whole body. Not only with the mind.

Healing Therapies?

For 25 years the Western world has seen a turmoil of 'alternative therapies' to heal ourselves and develop spiritually. They have all also been part of grappling for a new worldview and for a new understanding. Linking with the Source, divinity, the higher Self, or whatever we call the divine presence and cleaning, will change your thinking and raise your vibrations. But therapies do have a function as an introduction to learning what a spiritual worldview is like. But they are not a place to stop, just a place to pass through. The goal of life is to develop spiritually and thus become co-creative of evolution.

Most people in our society see the purpose of life as achieving success (a good job, recognition, wealth, etc.) and creating a loving sexual relationship with somebody of the other sex. Can this be spiritual? Western culture has not solved this issue. People spend their lives trying to get rich or in search of a satisfying sexual relationship with a partner. They make this the primary goal in life – a collective myth supported by the educational system and the media. But how can a goal in life only be associated with only the two lowest chakras; the root (material things) and the hara (sexual drive) chakras? Freeing up the hara chakra may make for better sex and more joy in life, but can still not be the goal in life. Staying with a partner and starting a spiritual journey together may be a better solution for the whole.

Conclusion

The goal for me must be the opening of the whole chakra structure and an integral practice both inner and outer so that divinity may be experienced, our hearts opened and our sacred Earth healed. Development of love or of higher consciousness (the upper chakras) has to be a more commendable goal for humankind. A simple life is the best and easiest way of reaching this and a life in community. On a more global level this purpose will lead to work for global justice and the continuation of evolution as described in module 1.

The burning desire to give back to life something unique that further enhances life dwells in every human heart. This is the desire that fuels the ecovillage movement and Gaia Education and which has inspired my husband to write his book, *Occupy World Street*. This is the same desire that was burning in Rashmi's heart.

Liberation

When I leave my prison,
I reach that state of liberation.
The mirror is disillusioned.
I merge with
time, space and the universe.
Then there is nothing
within or without,
you and I,
leaving or coming.
Clouds, sunshine, laughter, music,
love and joy.
All is being
and being is eternal.

Life – an Eternal Romance

Life is a dream
It is fire
It is energy
It is consciousness
It is experience
It is timeless.

Life is a burning desire
It is will to do the impossible
It is spirit for soaring high
It is chance to realise the potentials
It is power to create
It is struggle to reach the ultimate
It is drive to realise the dream.

Life is purpose
It is a cause
It is determination
It is courage
It is will
It is hope.

Life is an adventure
It is a battlefield
It is yearning to suffer
It is willingness to sacrifices
It is a kiss of blood
It is an invitation to death
It is beyond results
It is joy of burning.

Life is a passion
It is a challenge to nothingness
It is manifestation of excellence
It is liberation
It is romance
It is the essence of vision
It is a flame, in which dreams burn eternally
Life is a moment to be immortal.

Rashmi Mayur, 1995

References

Ari Ariyaratne; *Collected Works I-VII*; Sri Lanka.

Asger Lorentsen; 12 books in Danish, including *The Golden Circle*.

Eckhart Tolle; *The New Earth: Awakening to Your Life's Purpose*; Penguin, New York, 2005.

George D. Bond; *Buddhism at Work: Community Development, Social Empowerment and the Sarvodaya Movement*; Kumarian Press, Sterling VA, 2005.

Helena Norberg-Hodge; *Ancient Futures: Learning from Ladakh*; Sierra Club Books, San Francisco, 1991.

Kyriacos Markides; *The Magus of Strovolos: The Extraordinary World of a Spiritual Healer*; Penguin Books, London, 1985. See also *Fire in the Heart* and more.

Pracha Hutanuwatr; *Asian Future: Dialogues for Change*, vol 1 and 2; Zed Books, London, 2005.

Sahaj Marg, see www. SCRM.org for books, photos, talks, whispers and more.

Thomas Berry; *Dream of the Earth*; Sierra Club with The University of California Press, San Francisco, California, 1988.

Thomas Berry; *The Great Work: Our Way into the Future*; Bell Tower, 1999.

† Jes Bertelsen and his former tantra partner have written substantially about tantra in Danish. For literature in English see Stephen Wik's book, *Beyond Tantra* or works by Georg Feurstein or Daniel Odier)

Please see page 90 for Hildur Jackson's biography.

David Korten tells the story of the metamorphosis of the monarch caterpillar to a butterfly and introduces a powerful metaphor of cooperation rather than competition for the next phase of human evolution.

The Great Transformation: The Opportunity

David Korten

The light-skinned race will be given a choice between two roads. If they choose the right road, the seventh fire will light the eighth and final (eternal) fire of peace, love, and brotherhood. If they make the wrong choice, the destruction they brought with them will come back to them, causing much suffering, death and destruction.
Seven Fires of Prophecy of the Ojibwe people

We are now experiencing a moment of significance far beyond what any of us can imagine... The distorted dream of an industrial technological paradise is being replaced by the more viable dream of a mutually enhancing human presence within an ever-renewing organic-based Earth community.
Thomas Berry

Perhaps nature's most powerful metaphor for the Great Turning is the story of the metamorphosis of the monarch caterpillar to the monarch butterfly, popularized by evolution biologist Elisabet Sahtoutis. The caterpillar is a voracious consumer that devotes its life to gorging itself on nature's bounty. When it has had its fill, it fastens itself to a convenient twig and encloses itself in a chrysalis. Once snug inside, it undergoes a crisis as the structures of its cellular tissue begin to dissolve into an organic soup.

Yet guided by some deep inner wisdom, a number of *organizer cells* begin to rush around gathering other cells to form *imaginal buds*, initially independent multicellular structures that begin to give form to the organs of a new creature. Correctly perceiving a threat to the old order, but misdiagnosing the source, the caterpillar's still intact immune system attributes the threat to the imaginal buds and attacks them as alien intruders.

The imaginal buds prevail by linking up with one another in a cooperative effort that brings forth a new being of great beauty, wondrous possibility, and little identifiable resemblance to its progenitor. In its rebirth, the monarch butterfly lives lightly on Earth, serves the regeneration of life as a pollinator, and migrates thousands of miles to experience life's possibilities in ways the earthbound caterpillar could not imagine.

As the familiar cultural and institutional guideposts of Empire disintegrate around us, we humans stand on the threshold of a rebirth no less dramatic than that of the monarch caterpillar. The caterpillar's transformation is physical; the human transformation is institutional and cultural. Whereas the caterpillar faces a preordained outcome experienced by countless generations before it, we humans are path-breaking pioneers in uncharted territory. The rebirth is no wishful fantasy. It is already under way, motivated by a convergence of the imperatives and a spreading cultural and spiritual awakening of the higher orders of human consciousness.

The conditions of the human rebirth are likely to be traumatic and filled with a sense of loss, particularly for those of us who have enjoyed the indulgences of Empire's excess. Our pain, however, pales by comparison with the needless, unconscionable suffering endured for five millennia by those whose humanity and right to life Empire has cruelly denied. If we the privileged embrace the moment, rather than fight it, we can turn the tragedy into an opportunity to claim our humanity and the true prosperity, security, and meaning of community.

The cultural and spiritual awakening underlying the prospective human metamorphosis is driven by two encounters: one with the cultural diversity of humanity and the other with the limits of the planet's ecosystem.

The cultural and spiritual awakening underlying the prospective human metamorphosis is driven by two encounters: one with the cultural diversity of humanity and the other with the limits of the planet's ecosystem. A rapid increase in the frequency and depth of cross cultural exchange is awakening the species to culture as a human construct subject to intentional choice. The spreading failure of natural systems is creating an awareness of the interconnectedness of all life.

These encounters are bringing forth the higher and more democratic orders of human consciousness, expanding our sense of human possibility, and supporting the formation of powerful global social movements dedicated to birthing a new era of Earth community.

Extracted from The Great Turning – From Empire to Earth Community *by David C. Korten, Berrett-Koehler, 2007.*

Dr. David Korten is a respected author and a leading authority on the political and institutional consequences of economic globalization and the expansion of corporate power at the expense of democracy, equity, and environmental protection.

Joanna Macy and Dr. Chris Johnstone describes The Great Turning – one of the great stories of our time – when we move away from self-destructive political, economic and social systems to those that are natural and life sustaining.

The Great Turning

Joanna Macy and Chris Johnstone

In the Agricultural Revolution of ten thousand years ago, the domestication of plants and animals led to a radical shift in the way people lived. In the Industrial Revolution that began just a few hundred years ago, a similar dramatic transition took place. These weren't just changes in the small details of people's lives. The whole basis of society was transformed, including people's relationship with one another and with Earth.

Right now a shift of comparable scope and magnitude is occurring. It's been called the Ecological Revolution, the Sustainability Revolution, even the Necessary Revolution. This is our third story: we call it the Great Turning and see it as the essential adventure of our time. It involves the transition from a doomed economy of industrial growth to a life-sustaining society committed to the recovery of our world. This transition is already well under way.

In the early stages of major transitions, the initial activity might seem to exist only at the fringes. Yet when their time comes, ideas and behaviors become contagious: the more people pass on inspiring perspectives, the more these perspectives catch on. At a certain point, the balance tips and we reach critical mass. Viewpoints and practices that were once on the margins become the new main- stream.

In the story of the Great Turning, what's catching on is commitment to act for the sake of life on Earth as well as the vision, courage, and solidarity to do so. Social and technical innovations converge, mobilizing people's energy, attention, creativity, and determination, in what Paul Hawken describes as 'the largest social movement in history'. In his book *Blessed Unrest*, he writes: "I soon realised that my initial estimate of 100,000 organisations was off by at least a factor of ten, and I now believe there are over one – and maybe even two – million organisations working towards ecological sustainability and social justice."

Don't be surprised if you haven't read about this epic transition in major newspapers or seen it reported in other mainstream media. Their focus is usually trained on sudden, discrete events they can point their cameras at. Cultural shifts happen on a different level; they come into view only when we step back enough to see a bigger picture changing over time. A newspaper photograph viewed through a magnifying glass may appear only as tiny dots. When it seems as if our lives and choices are like those dots, it can be difficult to recognize their contribution to a bigger picture of change. We might need to train ourselves to see the larger pattern and recognize how the story of the Great Turning is happening in our time. Once seen, it becomes easier to recognize. And when we name it, this story becomes more real and familiar to us.

As an aid to appreciating the ways you may already be part of this story, we identify three dimensions of the Great Turning. They are mutually reinforcing and equally necessary. For convenience, we've labeled them as first, second, and third dimensions, but that is not to suggest any order of sequence or importance. We can start at others. It is for each of us to follow our own sense of rightness about where we feel called to act.

The First Dimension: Holding Actions

Holding Actions aim to hold back and slow down the damage being caused by the political economy of Business as Usual. The goal is to protect what is left of our natural life-support systems, rescuing what we can of our biodiversity, clean air and water, forests, and topsoil. Holding actions also counter the unraveling of our social fabric, caring for those who have been damaged and safeguarding communities against exploitation, war, starvation, and injustice. Holding actions defend our shared existence and the integrity of life on this, our planet home.

This dimension includes raising awareness of the damage being done, gathering evidence of and documenting the environmental, social, and health impacts of industrial growth. We need the work of scientists, campaigners, and journalists, revealing the links between pollution and rising childhood cancers; fossil fuel consumption and climate disturbance; the availability of cheap products and sweatshop working conditions. Unless these connections are clearly made, it is too easy to go on unconsciously contributing to the unraveling of our world. We become part of the story of the Great Turning when we increase our awareness, seek to learn more, and alert others to the issues we all face.

There are many ways we can act. We can choose to remove our support for behaviors and products we know to be part of the problem. Joining with others, we can add to the strength of campaigns, petitions, boycotts, rallies, legal proceedings, direct actions, and other forms of protest against practices that threaten our world. While holding actions can be frustrating when met with slow progress or defeat, they have also led to important victories. Areas of old-growth forests in Canada, the United States, Poland,

and Australia, for example, have been protected through determined and sustained activism.

Holding actions are essential; they save lives, they save species and ecosystems, they save some of the gene pool for future generations. But by themselves, they are not enough for the Great Turning to occur. For every acre of forest protected, many others are lost to logging or clearance. For every species brought back from the brink, others are lost to extinction. Vital as protest is, relying on it as a sole avenue of change can leave us battle weary or disillusioned. Along with stopping the damage, we need to replace or transform the systems that cause the harm. This is the work of the second dimension.

Vital as protest is, relying on it as a sole avenue of change can leave us battle weary or disillusioned. Along with stopping the damage, we need to replace or transform the systems that cause the harm.

The Second Dimension: Life-Sustaining Systems and Practices

If you look for it, you can find evidence that our civilization is being reinvented all around us. Previously accepted approaches to healthcare, business, education, agriculture, transport, communication, psychology, economics, and so many other areas are being questioned and transformed. This is the second strand of the Great Turning, and it involves a rethinking of the way we do things, as well as a creative redesign of the structures and systems that make up our society.

The financial crisis in 2008 caused many to start questioning our banking system. In a poll that year, over half those interviewed said interest rates used to be their main concern, but now they also considered other factors, such as where the money was invested and what it was doing. Alongside this shift in thinking, new types of banks, like Triodos Bank, are rewriting the rules of finance by operating on the model of 'triple return'. In this model investments bring not only financial return but also social and environmental benefits. The more people put their savings into this kind of investment, the more funds become available for enterprises that aim for greater benefits than just making money. This in turn fuels the development of a new economic sector based on the triple bot- tom line. These investments have proved to be remarkably stable at a time of economic turbulence, putting ethical banks in a strong financial position.

One area benefiting from such investment is the agricultural sector, which has seen a swing to environmentally and socially responsible practices. Concerned by the toxic effects of pesticides and other chemicals used in industrial farming, large numbers of people have switched to buying and eating organic produce. Fair-trade initiatives improve the working conditions of producers, while Community Supported Agriculture (CSA) and farmers' markets cut food miles by increasing the availability of local produce. In these and other areas, strong, green shoots are sprouting, as new organizational systems grow out of the visionary question, "Is there a better way to do things – one that brings benefits rather than causing harm?" I some areas, like green building, design principles that were considered on the fringe a few years ago are now finding widespread acceptance.

When we support and participate in these emerging strands of a life-sustaining culture, we become part of the Great Turning. Through our choices about how to travel, where to shop, what to buy, and how to save, we shape the development of this new economy. Social enterprises, micro-energy projects, community teach-ins, sustainable agriculture, and ethical financial systems all contribute to the rich patchwork quilt of a life-sustaining society. By themselves, however, they are not enough. These new structures won't take root and survive without deeply ingrained values to sustain them. This is the work of the third dimension of the Great Turning.

The Third Dimension: Shift in Consciousness

What inspires people to embark on projects or support campaigns that are not of immediate personal benefit? At the core of our consciousness is a wellspring of caring and compassion; this aspect of ourselves – which we might think of as our connected self – can be nurtured and developed. We can deepen our sense of belonging in the world. Like trees extending their root system, we can grow in connection, thus allowing ourselves to draw from a deeper pool of strength, accessing the courage and intelligence we so greatly need right now. This dimension of the Great Turning arises from shifts taking place in our hearts, our minds, and our views of reality. It involves insights and practices that resonate with venerable spiritual traditions, while in alignment with revolutionary new understandings from science.

A significant event in this part of the story is the Apollo 8 spaceflight of December 1968. Because of this mission to the moon, and the photos it produced, humanity had its first sighting of Earth as a whole. Twenty years earlier, the astronomer Sir Fred Hoyle had said: "Once a photograph of the Earth taken from the outside is available, a new idea as powerful as any in history will be let loose." Bill Anders, the astronaut who took those first photos, commented: "We came all this way to explore the moon and the most important thing is that we discovered the Earth."

We are among the first in human history to have had this remarkable view. It came at the same time as the development in science of a radical new understanding of how our world works. Looking at our planet as a whole, Gaia theory proposes that the Earth functions as a self-regulating living system.

During the past forty years, those Earth photos, along with Gaia theory and environmental challenges, have provoked the emergence of a new way of thinking about ourselves. No longer just citizens of this country or that, we are discovering a deeper collective identity. As many indigenous traditions have taught for generations, we are part of the Earth.

A shift in consciousness is taking place, as we move into a larger landscape of what we are. With this evolutionary jump comes a beautiful convergence of two areas previously thought to clash: science and spirituality. The awareness of a deeper unity connecting us lies at the heart of many spiritual traditions; insights from modern science point in a similar direction. We live at a time when a new view of reality is emerging, where spiritual insight and scientific discovery both contribute to our understanding of ourselves as intimately interwoven with our world.

We take part in this third dimension of the Great Turning when we pay attention to the inner frontier of change, to the personal and spiritual development that enhances our capacity and desire to act for our world. By strengthening our compassion, we give fuel to our courage and determination. By refreshing our sense of belonging in the world, we widen the web of relationships that nourishes us and protects us from burnout. In the past, changing the self and changing the world were often regarded as separate endeavours and viewed in either-or terms. But in the story of the Great Turning, they are recognized as mutually reinforcing and essential to one another (see box below).

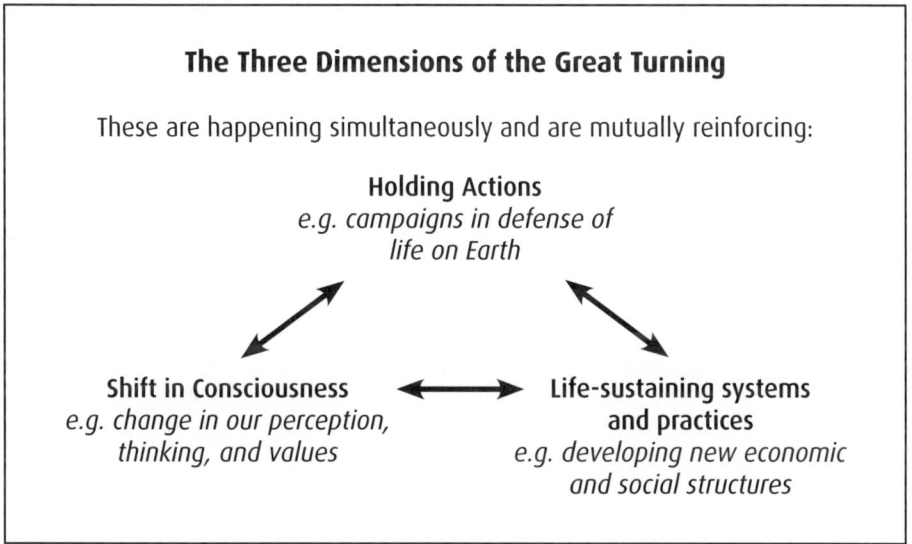

The Three Dimensions of the Great Turning

These are happening simultaneously and are mutually reinforcing:

Holding Actions
e.g. campaigns in defense of life on Earth

Shift in Consciousness
e.g. change in our perception, thinking, and values

Life-sustaining systems and practices
e.g. developing new economic and social structures

Future generations will look back at the time we are living in now. The kind of future they look from, and the story they tell about our period, will be shaped by choices we make in our lifetimes. The most telling choice of all may well be the story we live from and see ourselves participating in. It sets the context of our lives in a way that influences all our other decisions.

In choosing our story, we not only cast our vote of influence over the kind of world future generations inherit, but we also affect our own lives in the here and now. When we find a good story and fully give ourselves to it, that story can act through us, breathing new life into everything we do. When we move in a direction that touches our heart, we add to the momentum of deeper purpose that makes us feel more alive. A great story and a satisfying life share a vital element: a compelling plot that moves toward meaningful goals, where what is at stake is far larger than our personal gains and losses. The Great Turning is such a story.

This chapter is extracted from Active Hope: How To Face the Mess We're in without Going Crazy, *by Joanna Macy & Chris Johnstone, New World Library, 2012.*

Eco-philosopher Joanna Macy PhD, is a scholar of Buddhism, general systems theory, and deep ecology. A respected voice in the movements for peace, justice, and ecology, she interweaves her scholarship with five decades of activism. As the root teacher of the Work That Reconnects, she has created a ground-breaking theoretical framework for personal and social change, as well as a powerful workshop methodology for its application. Joanna is the author of numerous books. **www.joannamacy.net**

Chris Johnstone is a medical doctor, author, and coach. A former Senior Teaching Fellow at Bristol University Medical School, he trains health professionals in behavioral medicine and gives courses exploring the psychological dimensions of planetary crisis. Chris is known for his work pioneering the role of resilience training in promoting positive mental health, developing self-help resources and setting up the Bristol Happiness Lectures. He is author of *Find Your Power: A Toolkit for Resilience and Positive Change* (Permanent Publicaions, 2nd Ed, 2010) and co-presenter of *The Happiness Training Plan* CD (2010). See also **www.greatturningtimes.org**

Coming to us across twelve centuries, the Tibetan prophecy about the coming of the Shambhala warriors illustrates the challenges we face in the Great Turning and the strengths of compassion and wisdom that we can bring to it.

The Shambhala Prophecy

as told by
The Venerable Dugu Choegyal Rinpoche

A story that inspires both Joanna Macy and Chris Johnstone is a twelve-centuries-old prophecy from the Tibetan Buddhist tradition. The heroes of this story are called Shambhala warriors. Joanna and Chris explain that the term Shambhala warrior is a metaphor for the Buddhist figure of the bodhisattva, one who deeply understands the core teaching of the Lord Buddha. That central doctrine is the radical interdependence of all things. When taken seriously, this leads to the recognition that if one person has the capacity to be a bodhisattva, then all others do too.

Here is a particular version of the prophecy as it was given to Joanna by her dear friend and teacher Dugu Choegyal Rinpoche of the community of Tashi Jong in northwest India. Read it as if it were about you.

There comes a time when all life on Earth is in danger. At that time great powers have arisen, barbarian powers. And although they waste their wealth in preparations to annihilate one another, they have much in common. Among the things they have in common are weapons of unfathomable destructive power and technologies that lay waste to the world. It is just at this point in our history, when the future of all beings seems to hang by the frailest of threads, that the kingdom of Shambhala emerges.

You can't go there, because it is not a place. It exists in the hearts and minds of the Shambhala warriors. You can't tell whether someone is a Shambhala warrior just by looking at her or him, because these warriors wear no uniforms or insignia. They have no banners to identify whose side they're on, no barricades on which to climb to threaten the enemy or behind which to rest and regroup. They don't even have any home turf. The Shambhala warriors have only the terrain of the barbarian powers to move across and act on.

Now the time is coming when great courage is required of the Shambhala warriors – moral and physical courage. That is because they are going right into the heart of the barbarian powers to dismantle their weapons. They are going into the pits and citadels where the weapons are made and deployed, they are going into the corridors of power where the decisions are made. In this way they work to dismantle the weapons in every sense of the word.

The Shambhala warriors know these weapons can be dismantled because they are manomaya, which means 'mind-made'. They are made by the human mind and thus can be unmade by the human mind. The dangers facing us are not brought on us by some satanic deity or some evil extraterrestrial force, or by some unchangeable preordained fate. Rather, these dangers arise out of our relationships and habits, out of our priorities.

So, said Choegyal, now is the time for the Shambhala warriors to go into training. "How do they train?" Joanna asked. They train in the use of two implements, he said. Actually, he used the term weapons. "What are they?" Joanna asked, and he held up his hands the way the dancers hold up the ritual objects in the great lama dances of his people. "One," he said, "is compassion. The other is insight into the radical interdependence of all phenomena."

You need both. You need compassion because it provides the fuel to move you out to where you need to be and to do what you need to do. It means not being afraid of the suffering of your world, and when you're not afraid of the world's pain, then nothing can stop you.

But by itself that implement is very hot; it can burn you out. So you need the other tool, the insight into the radical interconnectivity of all that is. When you have that, then you know that this is not a battle between the good guys and the bad guys. You know that the line between good and evil runs through the landscape of every human heart. And you know that we are so interwoven in the web of life that even our smallest acts have repercussions that ripple through the whole web, beyond our capacity to see. But that is kind of cool, he said, even a little abstract. So you also need the heat of the compassion.

Scriptural source: the Kalachakra Tantra, 8th century AD.

This chapter is adapted from Active Hope: How To Face the Mess We're in without Going Crazy, *by Joanna Macy & Chris Johnstone, New World Library, 2012.*

MODULE 3
Reconnecting With Nature

Contents

Ancient Prophecies and The Vision Quest as a Path to Oneness

The Declaration of the Sacred Earth Gathering

Riding the Paradox: A Colourful Middle Way

The Bioregional Vision

The Path To Oneness: The Vision Quest

Voices of our Ancestors

Pathways to Integration:
Rediscovering the Song of the Earth

Seeding the Round Planet

Japanese Haiku

We did not come into world. We came out of it, like buds out of the branches and butterflies out of cocoons. We are a natural product of this earth, and if we turn out to be intelligent beings, then it can only be because we are fruits of an intelligent earth, which is nourished in turn by an intelligent system of energy.

Lyall Watson, Gift of Things Unknown

Hanne Marstrand Strong sets the scene by citing the prophecies ancient and modern by spiritual leaders who have been predicting our planetary and cultural crises. She suggests that we can all start the recovery process by personally reconnecting with the earth.

Ancient Prophecies and The Vision Quest as a Path to Oneness

Hanne Marstrand Strong

> *We must live in harmony with the natural world and recognize that excessive exploitation can only lead to our own destruction. We cannot trade the welfare of our future generations for profit now. We must abide by the Natural Law or be victim of its ultimate reality.*
> Tadodaho Leon Shenandoah,
> Grand Chief of the Six Nations Iroquois Confederacy

Many indigenous and non-indigenous cultures know that the root cause of environmental destruction stems from the disconnection between humans and nature. Humanity no longer lives in harmony with Natural Law that is represented by the elements; earth, air, water and fire. The Laws of Nature have supported all life since the beginning of time. Over the past hundred years or so, humankind has increasingly disconnected from the natural world. With the irresponsible use of technological advances and rapid unchecked industrialization, our separation from the natural world has accelerated and led us to our current global crisis and our very uncertain future. In this short period of time, we have destroyed what it took nature billions of years to create, causing major imbalances in our life giving elements. One result is the increased manifestation and magnitude of natural disasters around the world. We have become a culture incapable of feeling any emotional connection or relationship to our natural environment.

Clearly, this pattern is linked to a loss of understanding of the fundamental principles of interdependence; the wellbeing of all life-forms is interconnected. We have allowed ourselves to be distracted by meaningless pursuits. We have moved away from acknowledging and working towards our higher purpose and replaced it with insatiable consumerism and greed.

By contrast, indigenous cultures have lived a holistic existence of oneness – recognizing that there is no separation of body, mind, spirit and nature. Because of this profound understanding of life, these cultures lived in balance with the natural world. Through principles of oneness, a direct connection was maintained and viewed as a sacred state of existence between spirit, nature and humanity.

Prophecies Ancient and Contemporary

For thousands of years, humankind has been warned of impending disaster, one that threatens the very existence of life on earth. These warnings have come from prophets and people of various indigenous and religious traditions, many of which share similar scenarios on the fate of humankind. Human behavior, egoism and the focus on material gain have replaced moral and spiritual integrity and we are now witnessing the results. Perhaps the gravest result is climate change.

Many years ago, I visited the Prophecy Rock of the Hopi Nation in Arizona. According to Hopi tradition, the symbols on this rock warn of a time when this life will either be destroyed or restored to a paradise, depending on our actions as caretakers. In 1948 Hopi elders warned the world and later attended the Stockholm conference in 1972. In the early 1990s, the Kogi people, who live in the mountains of Columbia, came out for the first time to communicate with the outside world. They warned us all that we were killing the 'great mother' earth.

In eighth century Tibet, the prophecies of Padmasambhava (or Guru Rinpoche as he is also known) warned that certain conditions will come to pass in our time because of humanity's insatiable egoism and molestation of the natural elements. He knew of the changes to come. According to these prophecies, the celestial order is disrupted and the consequences are disastrous: plague, famine, chaos, and war lead to a panic that rages like wildfire; no rain falls in season, and when it does the valleys are flooded; drought, frost, and hail govern many unproductive years; earthquakes bring sudden floods while fire, storms, and tornadoes destroy entire cities; impostors and frauds cheat the people; fools preach the path to salvation; the advice of sycophants is followed; loquacity and eloquence pass as wisdom; the butcher and murderer become leaders of men; unscrupulous self-seekers rise to high position; behavior that was previously anathema becomes tolerated; good customs and habits are rejected and many disagreeable innovations corrupt the population; people die of starvation even as there is food to eat; the food itself becomes lifeless; and people are distracted by meaningless pursuits and bombarded with useless information that leads nowhere. Padmasambhava also said that one of the few remedies for this time would be to reforest the planet.

> *Humanity is in the midst of an epochal transition. The future of the planet and existence of all life is at a threshold. We humans live the destiny of our choices, and it is now every person's duty to aspire for the highest level of consciousness.*

Today, various contemporary scholars and scientists have delivered similar warning messages, exhorting humanity to change direction. Some of these clarion voices include Anne and Paul Ehrlich, Thomas Berry, Dennis Meadows, Donella Meadows, Maurice Strong, Rene Dubois, Barbara Ward, Dr. James Hansen, and Al Gore. Professor Schellenhuber has reported that climate change could reduce the population to one billion people. The United Nations has reported that agricultural production could be cut in half.

Why haven't we listened to these warnings that have been given throughout history?

Why are we willing participants in our own demise? Journalist Nicholas Kristof highlights a recent study on humanity's inability to deal with circumstances that we do not perceive as immediately threatening. Kristof explains that the human brain is prepared to grapple with imminent danger, but threats in the future do not activate our internal warning systems (*New York Times* Commentary Letters, 2009). Kristof writes that the most serious threats sneak under our brain's radar, and thus 'global warming doesn't ring our alarm bells'.

Humanity is in the midst of an epochal transition. The future of the planet and existence of all life is at a threshold. We humans live the destiny of our choices, and it is now every person's duty to aspire for the highest level of consciousness. This is the only way that we can create a hopeful future. What is demanded now is a vast shift in human consciousness.

The Vision Quest: Seeking Your Vision and Connecting with Your Spirit Helper or Guide

A powerful means to facilitate this needed shift of consciousness in individuals is the 'vision quest'. Traditional cultures have long recognized that each individual has a unique purpose in life, and the vision quest emerged as a practical way to discover one's unique potential; a rite of passage or way of transitioning from puberty to adulthood. 'Hanbleceya' or 'Crying for a Vision' is a Lakota rite of passage, connecting the seeker to the creator to discover one's purpose in life and to connect with a 'spirit guide' who remains active throughout the duration of life. The preparation for a vision quest begins 30 days prior; no alcohol, no intimacy and no bathing is permitted during this time and the seeker prepares an altar which consists of tobacco ties. The altar carries one's prayers and offers protection through the journey into the spirit world. A purification or 'Inipi' ceremony is then conducted in a sweat lodge, traditionally built by the seeker. Helpers go out ahead of time to prepare the sacred place where the seekers will go to seek a vision. Seekers carry their own pipes and are guided by a medicine man to an isolated place. This sacred place is often located on a high mountain, a bluff or in a pit dug into the ground. The seeker stays here for four days and four nights and prays for a vision. Visions often come to the seeker in the form of an animal, and dreams carry the most powerful messages. This ceremony is undertaken to undergo an experience of oneness.

Personal Experiences of Vision Quest

I started on the path of vision questing under the direction of my Cree teacher, Red Cloud in 1975. He introduced me to what would be my spiritual path over the course of my life. In the Fall of 1978, upon moving to Crestone, Colorado my path crossed with an elderly prophet named Glen Anderson. He appeared at my doorstep one day and said, "I have been waiting for you to arrive." He spent four days at my home and orally transmitted Crestone's destiny.

He relayed to me that the world spiritual traditions were to be represented here in Crestone, Colorado. Solar villages were to be built demonstrating to the world a life of simplicity, low consumption and sustainability. During our conversation he spoke about climate change and warned of extreme shifts in weather patterns. He said this community's main purpose was to help bring forth a new civilization of people who would live in peace and oneness. He predicted that people would come from all over the world to attain a higher consciousness and that many children would come for refuge and we must prepare to feed, cloth and guide them. He concluded, "This is why you are here." It was a lot to take in.

I felt the only way to confirm this prophecy was to go to the mountains for four days and four nights. During my meditation I received the message that Glen Anderson's transmission was the reason why I was brought to this sacred place. On the last night I heard a voice tell me, "You are not to leave this mountain until you have made the map." The question arose in my mind, "What map?". The reply was, "The map of the world traditions," referring to their placement in Crestone. I proceeded to visualize the map.

After my vision quest, everything started to unfold spontaneously. I felt as if the creator embraced this vision. I have subsequently committed my life to manifesting this interfaith community, following this experience in the mountains.

A few months after the vision quest, I was invited to a meeting in Hopi land, where I received confirmation from Hopi Elders that this vision was indeed meant to come into being. The Hopi elder Thomas Banyanca shared with me that this was also part of their Hopi prophecy to manifest on Hopi land. I found this very interesting, as the San Luis valley where Crestone is located was a place where indigenous cultures in the area would come to pray, and was once Hopi territory.

The following year I went to South Dakota, Crazy Horse's homeland. The Grandson of Crazy Horse's Medicine Man (Chips), Sam MovesCamp or Mato Blahitchya placed me in a hand dug pit in Bear Butte. He set up my protective altar around the periphery of where I sat, made up of hundreds of tobacco ties. During the four days and four nights, I sat with full exposure to the elements with only a blanket. As time passed without food or water, I became clearer and clearer, more connected to my surroundings. I began to feel very fluid, and my heart felt lighter and full of love. Tremendous joy arose, even though it was 110 degrees during the day and raining and thundering at night. I felt overwhelming respect for everything around me. There were no distractions to consume my mind, only nature and the living energies that surrounded me, the elements, the earth, sun and moon and the sky. I felt such gratitude, and realized that the elements were teaching me gratitude. My mind became ever clearer. It was as though nature became a spiritual filter system, filtering out negative emotions, and bringing body, mind and spirit into balance and in-tune with the natural surroundings. I felt strength, wisdom, and a direct connection to spirit.

At the end of the vision quest, the helpers came and took me back to the sweat lodge where I revealed all I had seen and heard to the Wicasa Wakan or medicine man, who interpreted the vision. I left the sweat lodge feeling much lighter of material attachments. I felt now that I was prepared and given the strength to carry out the work ahead of me. I had merged with Oneness.

I founded the Earth Restoration Corps (ERC) in 1990, which was later launched at the Rio Earth Summit in 1992. ERC is a training program designed to train and empower youth in restoring the planet. Young people are trained for sustainable jobs in Earth restoration, creating alternative livelihoods to alleviate youth unemployment by developing a new ecologically sensitive green economy. The primary technique used by ERC to facilitate transformation of consciousness and empower young people is the vision quest. Initially ERC participants spend twenty-four hours in nature on their first solo vision quest.

Re-establishing ways to connect to nature and experience oneness are vital to finding our place in the world.

The Ashaninka

ERC established a partnership with the Ashaninka in 2002. The Ashaninka are an indigenous people living in the rainforests of Peru and in the State of Acre Brazil. They have a long history of resistance, standing up to invaders since the time of the Inca empire until the rubber boom of the nineteenth century and resisting the encroaches of loggers from the 1980s to today, especially those on the Peruvian side of the border.

Moises and Benki Piyãko are two of ERC's Master Trainers and serve as part of the curriculum development team. They have been in the forefront of fending off the encroaching logging companies. Many Ashaninka have lost their lives trying to protect their lands. Both Benki and his older brother Moises found that it was futile to fight the loggers and mining companies and have implemented an alternative solution. They realize that most of the loggers from the surrounding communities live in poverty so this is a problem of economics, as these people struggle to feed their families. Benki and Moises introduced a brilliant solution... they started training surrounding indigenous and non-indigenous communities in sustainable livelihoods and sustainable forest management, creating green livelihoods (cottage industries) based on environmentally sound practices. Over time, they have essentially provided a whole new green economy of the Forest. Benki has now implemented 'the School of the Forest', training thousands of people annually.

Moises spends most of his time traveling to indigenous communities throughout Acre and the Amazon region, networking with other indigenous tribes, encouraging other tribes to return to the traditions that have been lost, and promotes the strengthening and revival of indigenous communities. Moises has introduced solidarity amongst 400 indigenous ethnic groups. He has strengthened and unified the voices of indigenous peoples throughout the region, in their ability to skilfully use non-violent means to fight off mining and logging companies.

Just as Moises and Benki Piyãko found their unique pathway to serve the preservation of their native Ashaninka culture, every human being has a unique role to play in serving the Earth. The vision quest is a remarkable gift tradition from indigenous cultures that can help anyone to discover her unique note in the symphony of life, and sing it out with abandon.

Hanne Marstrand Strong was born in Denmark and co-founded with her husband Maurice Strong the Manitou Foundation in Crestone, Colorado, a community of 34 spiritual centers. She was initiated by Native American spiritual leaders, and she organized the Sacred Earth Conference of international spiritual, religious and environmental leaders who presented 'The Sacred Earth Declaration' at the official opening of the Earth Summit in Rio de Janiero in 1992. During the Summit, Hanne also organized the 'Wisdom Keepers' who engaged in two weeks of ceremonial prayer and dialogue with indigenous leaders and religious leaders from around the globe. Hanne launched the Earth Restoration Corp (ERC) during the Wisdom Keepers gathering, and she regards education as crucial to provide young people with the necessary tools to restore themselves and the planet.

The Wisdom Keepers gathered together at the 1992 Rio Summit. There they co-authored a declaration that is both pithy and comprehensive and is also even more relevant today.

The Declaration of The Sacred Earth Gathering

by The Wisdom Keepers,
The Earth Summit, Rio De Janeiro, June 1992

The planet Earth is in peril as never before. With arrogance and presumption, humankind has disobeyed the laws of the Creator which are manifest in the divine natural order.

The crisis is global. It transcends all national, religious, cultural, social, political and economic boundaries. The ecological crisis is a symptom of the spiritual crisis of the human being, arising form ignorance. The responsibility of each human being today is to choose between the force of darkness and the force of light. We must therefore transform our attitudes and values, and adopt a renewed respect for the superior law of Divine Nature.

Nature does not depend on human beings and their technology. It is human beings who depend on Nature for survival. Individuals and governments need to evolve 'Earth Ethics' with a deeply spiritual orientation or the Earth will be cleansed.

We believe that the universe is sacred because all is one. We believe in the sanctity and the integrity of all life and life forms. We affirm the principles of peace and non-violence in governing human behavior towards one another and all life.

We view ecological disruption as violent intervention into the web of life. Genetic engineering threatens the very fabric of life. We urge governments, scientists and industry to refrain from rushing blindly into genetic manipulation.

We call upon all political leaders to keep a spiritual perspective when making decisions. All leaders must recognize the consequences of their actions for the coming generations.

We call upon our educators to motivate the people towards harmony

We believe in the sanctity and the integrity of all life and life forms. We affirm the principles of peace and non-violence in governing human behavior towards one another and all life.

with nature and peaceful coexistence with all living beings. Our youth and children must be prepared to assume their responsibilities as citizens of tomorrow's world.

We call upon our brothers and sisters around the world to recognize and curtail the impulses of greed, consumerism and disregard of natural laws. Our survival depends on developing the virtues of simple living and sufficiency, love and compassion with wisdom.

We stress the importance of respecting all spiritual and cultural traditions. We stand for preservation of the habitats and lifestyles of indigenous people and urge restraint from disrupting their communion with nature.

The World Community must act speedily with vision and resolution preserve the Earth, Nature and humanity from disaster. The time to act is now. Now or never.

Pracha Hutanuwatr & Jane Rasbash frame our relationship with nature and our natural resources in Buddhist and Taoist values and highlight how alien they are to western consumerism.

Riding the Paradox: A Colourful Middle Way

Pracha Hutanuwatr and Jane Rasbash

When we facilitate the worldview sessions for Buddhists and Chinese participants of Ecovillage Design Education (EDE) program, we borrow basic frames of thought from Buddhism and Taoism. This seems to get the message of the current paradigm shift across quite smoothly and deeply among participants. They complement the stories of the paradigm shift of new science emerging in the West that we also present to the groups. It is, however the Yin Yang framework from Tao philosophy and the middle way framework from Buddhism that illuminate, inspire and empower Asian participants as they realize the wisdom of their cultural roots.

Some frames of thought we work with to portray these Eastern approaches are:

The Named and the Unnamed
The Known and the Unknown
Using Nature and the Respect for Nature
Individual and Community
Competition and Cooperation
Linear Progress and Cyclical traditions
Satisfaction of Need and Reduction of Need

This chapter will discuss each of these lines briefly, showing the wisdom and knowledge of these Eastern traditions. We suggest that these not only inspire Asian participants to consider shifting their worldview to a more holistic approach, but also offer fertile material for EDE participants studying around the world.

Named and Unnamed

The first chapter in Tao Te Ching says:

> *The Tao that can be spoken of is not the eternal Tao;*
> *The name that can be named is not the eternal name.*
>
> *The nameless was the beginning of heaven and earth;*
> *The named was the mother of ten thousands things...*
>
> *These two are the same*
> *But diverge in name as they issue forth.*
> *Being the same they are called mysteries,*
> *Mystery upon mystery –*
> *The gateway of all mysteries.*

In Buddhism we are taught to live in double realities; the conventional truth (*samatisacca*) and the ultimate truth (*paramtthacacca*). The former are the names, concepts and values we give to things and the latter is the unnamed reality behind those names. Right and wrong, good and evil, whole and part, you and I, heaven and earth, success and failure, fame and defame, brother and sister, man and woman belong to the first category which are important and very useful in our daily life and in setting up social structures and conventions. Both Buddhism and Taoism teach us that this category is half of the truth or just a dimension of the total reality. We need to be aware of and live in the other dimension that has no name, no right or wrong, no good nor evil, no success nor failure, no I nor mine, no brother nor sister, no mother nor son...

In real life situations both dimensions exist together as a whole and can't be categorized. In modern society we are taught to live only in the conventional truth mode and this makes us lose our balance. For followers of Buddhism and the Tao, taking this part of reality too seriously is the source of all the other problems we are facing.

The Known and the Unknown

Both traditions tell us not to be over-confident about what we know. What is known to human beings is a very small part of the whole reality. We know the world through our senses and our senses are very limited in the unknowable universe. Human beings have been unsuccessfully trying to break this limit throughout human history despite all the tools we have created to extend the capacity of our senses of perception. The more we know the more we know that we don't know. Let's take our own bodies, even with the most advanced scientific knowledge, most sicknesses are still unexplainable to even the best doctors apart from some basic mechanical aspects. Not to mention our own minds, most of us don't know for sure why we are often unhappy. The

> *What is known to human beings is a very small part of the whole reality... The more we know the more we know that we don't know.*

tragedy is that most modern people are taught to believe that we know a lot or that the world is knowable. Tao and Buddhist teaching tell us to be cautious. What we know is what we interpret through our senses and pre-perceived memories. The Tao Te Ching says that "Knowing that you don't know is strength. Not knowing that you don't know is peril." And Buddhists says that 'the finger pointing to the moon is not the moon'. The Buddha himself teaches us not to believe in what he said uncritically.

Use of Nature and Respect for Nature

We are taught to apply the above assumptions regarding the unknowable nature. Of course as human beings, like other living beings, we are part of nature and have to make use of nature for our survival. However we should use nature with care and only take what we need. "Earth provides enough to satisfy everyone's need, but not everyone's greed," as Gandhi says. Both traditions teach us to be contented and happy with a simple way of life. The highest symbol of our Thai civilization is a small thatched roof hut in a forest monastery where a monk or nun lives and meditates with one meal a day that they receive from their morning alms round. They own only three robes. Everything else belongs to the community. The deep ecology notion of intrinsic value of all beings is not possible without a strong culture of contentment.

When we look with critical awareness it is not surprising that American development experts coming into Thailand in the 1950s and 1960s advised the dictator prime minister of that time to stop the Buddhist monks teaching contentment to Thai people. Given its rich natural resources, the Thai people were seen to be lazy, even by Thai elites who had been educated in the West. Now after 50 years of Americanised development, the Thai forests are reduced from 50% to less than 10%; most rivers are polluted and hundreds of thousands of fish have on occasion suddenly died all over the country; agriculture has turned into chemical and mechanized agri-business; most farmers are in debt; shopping malls in the big cities have become the new temples where families spend money on consumer goods rather than time meditating; and consumerism replaces the culture of contentment.

Of course, Thai people have become less happy and the country is much, much poorer in terms of natural resources and there is a widening gap between the rich and the poor. These phenomena is spreading to all countries in Asia now. Naturally, there are resistance and alternative movements in every country such as the Gandhian movement in India, the Saravodaya movement in Sri Lanka, Spirit in Education Movement in Thailand, and Prasantran Movement in Indonesia.

Individual and Community

The present paradigm overemphasizing the named and the known also create tremendous disaster to human individuals with the misinterpreted idea of self-interest and individualism. The individual becomes the crux of all things that matter for modern people at the expense of family, community and other social ties. From the Buddhist point of view individuality is valid to a point, however when it is over emphasized it becomes individualism. This makes the individual lose their sense of belonging to a community. This is a source of modern alienation and loneliness. Even with all the wealth, power and recognition that one earns through hard work life is still meaningless if you can't develop a deep relationship with anybody. The worst is that you do not even have a deep relationship with yourself because we are taught that the most important aims in life are wealth, power and recognition. From the Buddhist and Taoist points of view, this aspect of alienation is even more basic as you lose touch with the core of being human.

Of course it is true that traditional society, which defines who you are by where you belong, is not always rosy. There are many stories of collective tyranny over the individual both in the east and the west. Hence there has been a search for humanism since renaissance times. When the pendulum swings too far, however, we lose the balance between being individual and being part of a community.

From the Buddhist point of view individuality is valid to a point, however when it is over emphasized it becomes individualism. This makes the individual lose their sense of belonging to a community.

Competition and Cooperation

With the wrong kind of education and mad mass media advertising, we are taught to compete with each other in a materialistic way. Since kindergarten children are indoctrinated to feel they are not good enough as themselves. They have to strive to be someone else all the time. They have to do 'this' better and 'that' better to be good enough to be loved. Being wounded this way since early youth and being bombarded throughout adolescence by advertisements and education systems to support this inferiority ensures that when children grow up they become adults who are discontented with themselves. They have been programmed to believe that they must buy this or that to be complete and whole... and then another thing and so on it goes. A deep sense of not being good enough is inflicted into the soul which is difficult to heal with therapy. The modern world is full of these wounded souls running the machinery of our society and business circles with endless craving for more wealth, power, recognition, and sensual pleasure without knowing why he or she does so. Hence the desperate need for rebuilding community in modern society.

This is why the EDE is very relevant everywhere. We find modern men and women yearning for a sense of belonging to a community. Yet once he or she finds a community it is so difficult to live in it. The need to reeducate oneself for interpersonal relationships is tremendous. The deeper core essence of compassion needs to be rekindled. This cannot be learned in the conventional classroom because this kind of knowledge cannot be learned through lectures or reading books. It is not convergent knowledge (ref. E. F. Schumacher) like making a computer or aircraft.

Linear Progress and Cyclical Traditions

From a Buddhist and Taoist point of view, modern people are also confused about the idea of progress side by side with the idea of knowledge. The knowledge that allows us to make a computer or aircraft or atomic bomb is completely different from the knowledge that allows us to have a deep relationship with oneself and others or the knowledge of creating a democratic, just and sustainable society. The former you can accumulate, the latter you cannot. A Buddhist challenge is to say that modern society is not more democratic than the Buddhist sangha 2,500 years ago. Of course in some aspects we are grateful for progress – to have your tooth removed without pain is wonderful – but this depends on the accumulation of technical knowledge in dentistry. For a dental faculty to produce good dentists that care for their clients more than money, however, demands a different kind of knowledge and educational approach.

Traditionally both Buddhists and Taoists and other Asian traditions see history as a cyclical process rather than linear progress. The quality of the heart of the people of each era determines the quality of the society of that era. With the present ethos of promoting greed and violence in modern society there is no hope for democracy, justice and sustainability.

Satisfaction of Need and Reduction of Need

With the misunderstanding of 'progress' modern society also misunderstands happiness. The assumption is that the more you satisfy your craving for wealth, power, recognition, and sensual pleasure, the happier you are. Buddhist and Taoist masters teach the opposite. The Western green approach of differentiating needs and wants is a step in the right direction but does not go far enough because the definition of needs is slippery. In Buddhism even basic needs can be reduced. For Samdhong Rinpoche, the most prominent Buddhist thinker and former prime minister of Tibet, the route of modern evil is to produce surplus by modern technologies. Because you have to stimulate demand (greed) to meet oversupply. Hence modern needs are extended endlessly.

For the reconstruction of healthy communities and a healthier planet, we need healthy individuals who greatly reduce their material needs. We don't have to live a monastic life like a monk or nun who has three robes, one begging bowl and a hut little bigger than their body size but we can take this as a role model. Modern men and women with multiple wardrobes of clothes and other possessions are excessive. Buddhadasa Bhikku, a prominent Thai monk and Buddhist thinker, defines development as 'messiness and madness'.

In our trainings to raise awareness around greed and consumer monoculture, we took Buddhist monks in their yellow robes to do a vision quest in one of the biggest and busiest shopping malls of Bangkok. Monks sat and meditated before walking mindfully and slowly in and inside the mall. They had money but were not allowed to buy anything or talk to anyone. The instruction is to observe what happens in the mind when seeing all the stuff in the shopping mall. Deep ecologists Elias Amidon and Elizabeth Roberts designed this session when they introduced deep ecology in Thailand and Southeast Asia in the 1990s. It has been used regularly since then with many different groups and is effective in bringing awareness of how consumer culture stimulates greed in society.

In conclusion, we believe that the basic values of these two spiritual traditions are in common with other contemplative traditions around the world. The lists of these pairs of paradox are endless. The middle way is not a grey compromise of black and white, it is the vibrant and colourful coming together of yin and yang. It demands a deeper human potential and wisdom to ride these pairs of paradox which go far beyond the command of the intellect. There is no doubt that intellectual understanding in the right direction can help a lot but that will never be enough. What is needed is a new kind of education that nurtures and nourishes the healthier qualities of being human: compassion, loving kindness, wisdom of the oneness of all things, the spirit of non-harm and service to others. This means a transformation of consciousness from self-centredness to more ecological centredness. From ego-self to eco-self. This kind of education is not completely new, it can be informed by all contemplative traditions. Shamanism and some modern schools of psychotherapy such as the Jungian schools can be complementary. The EDE at its best is a great introduction and inspiration of these healthy life-long practices for the participants, their communities and our Planet – Mother Earth – Gaia.

The middle way is not a grey compromise of black and white, it is the vibrant and colourful coming together of yin and yang. It demands a deeper human potential and wisdom to ride these pairs of paradox which go far beyond the command of the intellect.

Pracha Hutanuwatr is an author and scholar of globalization and socially engaged Buddhism from Thailand. He is Director of the Wongsanit Ashram near Bangkok; Deputy Director of Santi Pracha Dhamma Institute; Program Director for Grassroots Leadership Training; and a board member of the Spirit in Education Movement. Pracha has lectured and led workshops internationally. His latest book is *Asian Futures: Dialogues for Change*, co-authored with Ramu Manivannan, Zed Books, 2005.

Jane Rasbash works in sustainable development using an empowerment and engaged spirituality approach. She lives in Findhorn Ecovillage and has taught on EDEs in Findhorn, Sieben Linden and Thailand. She is a Board Member of Gaia Education. Jane is currently serving as strategic and technical consultant for community, sustainable education and rights based projects in Burma, Thailand and more recently with GEN in some African countries. Jane is a life coach who mentors leaders in the south often working in incredibly challenging situations. She is passionate about Traditional Village as Ecovillage.

Gene Marshall explains why bioregionalism is a key to ecological and peaceful planetary co-existence.

The Bioregional Vision

Gene Marshall

The term 'bioregional' points to human beings living in committed relationships with local regions of the natural planet. A person or a group enters into the bioregional family of society-builders when that person or group subscribes wholeheartedly to these simple statements:

Earth is my home: I am an Earthling.

A continent of Earth is my home.

A region of Earth is my home.

This fresh sense of home is simple, but it has implications:

The United States, Canada, or Mexico or some other nation is not my home; it is just my nation.

My state or province is not my home; it is just my state or province.

My zip-code district is not my home; it is just my zip-code district.

Western civilization is not my home; no civilization is my home; it is just my civilization.

If you are a tribal person, your tribe is not your home; it is just your tribe.

Tribal people are bioregionalists, compared to 'civilized' people, because they have traditionally honoured all the living and inanimate beings in a specific region of the planet. They are not bioregionalists, however, because they are tribal and tribalism divides and separates members from non-members. When we apply the bioregional sense of home to envisioning the future of human society, we do not see tribes or civilizations. We see a planetary confederation of semi-autonomous Earth-regions.

When we apply the bioregional sense of home to envisioning the future of political and economic systems, we do not see a global economy ruled by wealth and unrefereed by local regions of people. We see popular consensus-building beginning in each local region and extending into an Earth-sensitive governance of the entire economic playing field for all players across the whole planet.

When we apply the bioregional sense of home to envisioning the future of human cultures, we do not see planet-wide uniformities conceived by product advertisers. We see local families of plants, animals, and humans forming unique expressions of aliveness in each region of the planet.

Such a vision is basically simple, but it has far-reaching implications. It means shutting down in our own minds the dream of building a better civilization or the dream of returning to a new sort of tribal life. It means dreaming a new dream. This new dream is not something grandly idealistic; it is a realistic direction for avoiding untenable ecological disaster. While we may learn many lessons from a thousand centuries of tribal society and sixty centuries of civilization, we must now create something new. We must see both civilizational hierarchy and tribal intimacy as obsolete patterns of living that are no longer appropriate for the real situation in which we dwell. We must dream a new dream. Bioregionalism is a name for that new dream.

While we may learn many lessons from a thousand centuries of tribal society and sixty centuries of civilization, we must now create something new.

Gene Marshall has a long history of participation in Christian renewal and interreligious dialogue. For the last 30 years, he has been doing full-time research and training for Realistic Living, a nonprofit organization for religious and ethical research.
www.realisticliving.org

Dhyani Ywahoo of the Eastern Tsalagi people describes the intimate relationship between the natural world through the Five Directions and the basis of wisdom.

VOICES OF OUR ANCESTORS

Dhyani Ywahoo

Each of us has a song in our heart. Through thought, through action, each one is creating vibration in the atmosphere. When we think about the dancing atoms that build and sustain the forms of life, we can see ourselves forever in the dance. Everything is vibration. Our action rings out in many dimensions and in that way our thoughts and actions return to us. We may call it karma or destiny; we are living the result of our thoughts and their various overtones interacting with other thoughts. Just as a pebble of intention dropped into a still pond ripples out in every direction, so the appearance and phenomena of our lives and our world are ripples on the serene surface of universal mind.

As the world spins we learn from the Adawees, wise protectors of the directions, guardians of mind's gates, giving form to the formless. Their wisdom expressed through the five wisdoms: the wisdom of the sphere of existence, recognizing the fundamental unity of things notwithstanding differences in external aspect; the wisdom that causes works to succeed; the wisdom that distinguishes particulars, the wisdom that equalizes, makes known common factors; the mirror-like wisdom, wisdom that reflects things as they are.

We recognize the five wisdoms in the stream of wisdom ever flowing within us, perceived in terms of the Medicine Wheel and the sacred directions and in relation to the five organ systems and the five elements. As we have a clear perception of how our minds express wisdom, how we apply it, we have greater choice about how we honor the movement of life within ourselves.

In the frozen waters of the North the mirror-like wisdom reflects our action, its causes, and its ripples in the stream of time and mind and relationship. Guardian of the East, Nutawa, Sunlight Arising, shines wisdom of inspiration, existence; one realizes one is alive, one's gifts becomes apparent. From the South the grandmothers dance forth their seed-baskets of good cause, generation, wisdom to succeed, bringing things to fruition. In the West we understand patterns, the wisdom of particulars, as Great Bear dances on ignorance: "Transform, transmute, bring forth what is beneficial in the spiral dance of life." In the center, the hub whence all experience arises, is the wisdom that equalizes.

Always we must come around the circle to find the harmony in ourselves. It is never really lost; we have only to accept it and let ourselves resonate with the whole universe. That resonance is affirmation of oneness. Our consciousness is shaping what is around us. We have called one another together as a family of humanity, and our duty as human beings is to see beauty, to sing, to be joyful, to work the land and to share the abundance. You have learned some great wisdom? Pass it along. You have a gift with sound, you have a gift with herbs? Share it. It is the weaving of the overtones of consciousness, of our minds, that repairs the sacred tapestry of life. So it is for each of us and all of us to make the fabric whole again – and the weaving begins in the heart, realizing unity within. It is simple and it calls upon deep discipline, because this body is the instrument and your thoughts are the musician.

The Tsalagi people say that we, together, are making a world, our thoughts and our interactions very much shaping the elemental forces of life. And with right thought and right action, especially the understanding of our purpose, our gift in this time, we may bring forth abundance and a great peace through peaceful understanding in ourselves and rekindling the sacred Wisdom Fire.

From Voices of Our Ancestors: Cherokee Teachings from the Wisdom Fire, *Dhyani Ywahoo, Shambala Books (1987).*

We have called one another together as a family of humanity, and our duty as human beings is to see beauty, to sing, to be joyful, to work the land and to share the abundance.

Dhyani is a member of the traditional Etowah Band or the Eastern Tsalagi (Cherokee) Nation. She is active in mediation circles.

Maddy Harland describes practical ways in which we can reconnect with nature as individuals and groups, sharing the joy of our exquisite natural world and celebrating life on our living planet.

Pathways to Integration: Rediscovering the Song of the Earth

Maddy Harland

It isn't a coincidence that we speak about 'earthing' or grounding ourselves to establish or maintain a sense of stability or a still centre in our frenetic modern world. Often on completing a meditation practice, we ground ourselves by visualising our 'roots' going deep into the Earth to echo the lotus plant, a symbol of spiritual elevation, with its flower above water and its roots deep in the mud of the pond, spanning three physical mediums.

Connecting with nature can be a deeply healing experience. It can stimulate positive experiences, make us happier people, bond a group together, and prompt a shift in worldview, from separation to integration. I remember once walking on the South Downs in England one sunny day and becoming aware that everything in the landscape was infused with sparkling, dancing energy. It emanated from every blade of grass, every stone, the sky, the clouds and even the lone figure walking (me!). I became strongly aware that the landscape and myself were made of the same essence behind the forms and that I was part of a greater unity. It was blissful.

The great mystical poet Rumi wrote:

> *I am the dust in the sunlight, I am the ball of the sun…*
>
> *I am the mist of morning, the breath of evening…*
> *I am the spark in the stone, the gleam of gold in the metal…*
> *The rose and the nightingale drunk with its fragrance.*
>
> *I am the chain of being, the circle of the spheres,*
> *The scale of creation, the rise and the fall.*

I am what is and is not...

I am the soul in all.

There are many ways in which we can connect with nature and experience the intelligence and inter-connected web of our beautiful planet Earth. Working together in groups is especially fertile ground as not only does it bring a greater connectivity with nature, it also can enhance the group's sense of identity. We do not always have to initiate formal group sessions either. Cooking a meal over an open fire, playing music together with household objects as percussion instruments, making art from local materials, sleeping under the stars, or swimming in wild water can all engender this kind of awareness. If the group is to be together for four weeks, creating a small but biodiverse salad garden is another technique. Gardening opens our awareness to the changing elements, microclimates, the quality of water, the seasons, the local wildlife and the process of growth. There is little more miraculous than growing food from seed, and to appreciate the pattern inherent in the form and be an agent of its fulfilment by good husbandry. Planting seeds can also be a deeply symbolic act as we can plant our intentions as individuals or as a group as we sow.

There is little more miraculous than growing food from seed, and to appreciate the pattern inherent in the form.

Breathing With Nature

Another technique for a group to share is to go out individually and silently into a natural area, agreeing to meet again in about 20 minutes. Treat it like a meditation and encourage everyone to slow down and be silent. As you walk, sense what you are drawn to. When you are attracted to a plant or tree, ask for permission to visit it. If you do not feel invited, move on. If you feel you have permission, sit with the plant and explore it with your five senses. Breathe with the plant, exchanging gases. Imagine how the plant is providing you with oxygen and you are providing the plant with carbon dioxide. Both of you need each other. Visualise symbiosis. Imagine how long this planet's ecosystems has depended on this exchange of gases. When you have finished, express your gratitude to the plant or tree and the natural area. When you gather together again, ask the group to share their experiences and how it felt. There is no need to interpret, explain or compare experiences. Just share them without judgment. Thank each other for doing this activity together.

Celebration

Whatever our cultural background or beliefs, making a sacred space can be both fun and hold a deep reverence. It can be a personal celebration of nature and natural forms and it can also be a still space in the main teaching area of a group. This can either be done together as a group endeavour or the facilitators can ask members of the group to add to the space as and when appropriate. A candle or incense can be lit every day and the group can collectively place positive qualities or intentions into the process.

How this space is used, however, needs to be in relation to the group and its culture. We are here to encourage each other but not to impose our individual culture or beliefs. When making a nature shrine, consider using objects that symbolise the four elements – water, air, earth and fire – and ideally place them in relation to the four directions. This practice is shared by many indigenous cultures, including the Celts and the Native Americans. It enables people to deepen their orientation. Where does the sun rise and set? Where do the winds and rains predominantly come from? Where does the moon rise and set? You may also like to include the 'fifth' direction – the inner.

This process of orientation with the four directions can be expanded. Times of transition – between night and day at dawn or dusk or the rising of the new moon or full moon – can be powerful times of connection. Take a walk at these times, listening to the nocturnal birds and animals waking up or going to roost and returning to their places of rest, observe the opening and closing of flowers, the changing light in the sky, the dayblind stars and planets... Theses are times of inner expansion and opportunities for a freshness of vision. Savour them.

Dream Recall

Another way of deepening connect with the natural world is to practice dream recall. Start a dream diary and note down your dreams immediately when you wake up. Often your unconscious mind will respond by offering you deeper, more vivid dreams. Add the dimension of a morning group session briefly recalling any significant dreams and you will have a dynamic route into the collective imagination of the group. Archetypal symbols and shared inner landscape may evolve. The trick is not to be too analytical, to try not to interpret everything symbolically, but to flow with the dream imagery and allow it to settle in your awareness during your waking life, just as you would a landscape or beautiful garden. From this reflective state arises insights, some personal but some also more inter-personal, even universal.

The Biotime or 'Phenomenal' Diary

Max Lindegger, the ecovillage designer and permaculture specialist, told me a story about the farmers in Northern Thailand. "One muggy afternoon, I expected it to rain any minute. I noted that our farmer was watering his vegetable seedlings. "Why bother?" crossed my mind, "Nature will do it very soon!" Next door the farmer lit the stubble. I wondered why? "It would rain soon anyway and kill the fire!" But it did not rain! How did the farmers know?"

Tang, one of our students told me the following: "Because farmers observe all the time and have done so for many years (and many generations) they note even slight changes. Ants will leave the soil, looking for higher ground

"Before the Tsunami of 26th December 2004 hit a small isolated island, the local Shong people left for higher and safe ground. None died. When asked how they knew a disaster was about to hit, their response was simply, 'Our ancestors told us.'"

and even shifting their eggs before heavy rain. A few hours before rain, insects will be attracted by light. Farmers make use of these phenomena, and set lights on ponds to attract insects to where fish can eat them. Farmers will listen to the sound of crickets as their sound changes when it is about to rain. Frogs will start to 'sing' before the rain arrives. When the seedpods of Tamarind curl up tight and are crunchy, cold weather is not far away. Earthworms will rise to the surface and often die and this is an indicator of cold weather coming.

"'Phenomenal' calendars have been part of many cultures. I remember a story from Native Americans that described 'the time to plant corn is when the size of oak leaves is equal to the size of a squirrel's ear'. This is more accurate information than a planting date as it is likely to relate to soil moisture and soil temperature.

"Many of these observations can be scientifically proven but some may simply be beyond science. Before the Tsunami of 26th December 2004 hit a small isolated island, the local Shong people left for higher and safe ground. None died. When asked how they knew a disaster was about to hit, their response was simply, 'Our ancestors told us.'"

It is useful to note the subtle changes in the natural world around us and it deepens our connection as well as provides a useful record in a world of changing climate. Acquire a book with blank pages (enough to have at least half or a whole page for every day of the year). Start on January 1 and write the day (but not the year) in its allocated place. Then simply start a diary of biological events. This can be when a species of migrating birds like swallows arrive or leave, when you see your first favourite wildflower, when a specific fruit tree come into leaf, when your poultry lays the first egg of the year, or when the fish begin to spawn... Allocate a colour for each year and make a key at the beginning. That way you can identify what year relates to each entry and begin to make annual comparisons on seasonal changes. You will be surprised by the results.

Tradition tells us that walking long distances to a sacred site can have a transformative effect on the pilgrim. We put aside our daily cares and focus entirely on a special journey.

Modern Pilgrimage

Tradition tells us that walking long distances to a sacred site can have a transformative effect on the pilgrim. We put aside our daily cares and focus entirely on a special journey. His Holiness The Gyalwang Drukpa is the head of the Drukpa lineage, the Buddhist sect of Ladakh in the Himalayas, has begun a new tradition that marries pilgrimage with environmental activism and a care for the Earth and His people. This synthesis of concerns is part of an emerging theme across continents. Ladakh itself is part of what is called the planet's 'third pole', a vast glacial region now devastated by climate chaos associated with global warming. This region is inhabited by three million people and holds a third of the world's fresh water. The seasons are unpredictable, hit by blizzards in summer and terrifying cloud bursts in which two inches of rain falls in 60 seconds. A brittle, arid landscape, it is littered by throwaway plastics that pollute the watercourse from which people drink.

To the local people, climate chaos is caused by the industrialised world. They are also unaware of how toxic degraded plastics are. To bring knowledge to the region, HH The Gyalwang Drukpa walks 450 miles from Sikkim through Nepal to Ladakh with 700 Buddhist monks, nuns and lay people every year. They pick up half a ton of plastic waste on the journey, hold workshops, educate the local people about climate change, the environment, organise mass tree plantings and practice Buddhist rituals together. People are encouraged to act and not to feel that they are powerless third class peasants at the mercy of a western culture that insidiously colonises the mind and culture through brand advertising. He also teaches the nuns Kung Fu and promotes them as equals, often unusual in a traditional society. Walking is a means of spreading knowledge, sharing concerns, building community and connecting deeply with this extraordinarily beautiful region of the Earth.

HH The Gyalwang Drukpa has made a film about this epic journey called 'Pad Yatra', a film that Joanna Macy and Chris Johnstone would call 'a story of our time'. At the centre of film is The Gyalwang Drukpa's teaching that reconnecting with Nature, environmental awareness and care for the Earth are at the core of his spiritual practice. It is the act of walking, of pilgrimage, that deepens these understandings and also brings great joy for we live in beautiful world, teeming with sentient life.

Walking in the Footsteps of our Ancestors

How can we reconnect with the Earth, wherever we live, to develop that sense of reverence, joy and commitment to the creation of a better world? I believe that the simple act of walking can offer us a profoundly powerful route to expanding our consciousness and appreciating what we have in our own locality.

Last summer I went walking along part the Ridgeway, Britain's oldest road, that runs through central southern England, through the wooded hills and valleys of the Chilterns to the north Wessex Downs, rich in wildlife found in chalk grassland habitats, and down into the World Heritage Site of Avebury. The Ridgeway follows an ancient route over the high ground used since prehistoric times by travellers, drovers and soldiers. Close by in the wide rolling landscape are many archaeological treasures: Neolithic and Bronze Age barrows, Iron Age forts, examples of ancient strip farming and the figures of white horses cut into the chalk. It is a landscape alive with the past, a feast of prehistory, Albion at its most monumental.

This was no ordinary hike. I had joined a group of people who chose to walk and live 'in community', sharing route and food, mainly rough camping together along the way. Arriving by train in Wiltshire in the Vale of the White Horse, where we were to begin our journey, we could only take what we could carry there and that had to include tent, sleeping bag and mat. By necessity, we had to live simply. Each day we walked, sometimes in community, sometimes alone, but always high up in a landscape offering wonderful panoramic perspectives. Graham Joyce, an archaeologist by training who had organised the walk, taught us the subtle details of prehistoric forms, of barrow and dyke, and to see traces of ancient humanity's agriculture, still evident on the chalky hills.

The group itself was rich with experience: there were artists, musicians, bushcrafters, herbalists, a teacher, a yurtmaker, yoga practitioners, permaculturists, long distance walkers... The camp evolved its own rhythm as we walked the hills and valleys each day and we learnt to care for each other, everyone generous with the daily tasks. The warmth of this temporary community was intoxicating. No egos jostled for space, each allowing the others to be nurtured and thrive. It sounds idyllic, but for those brief nine days it was ... and we all laughed more than we had for years.

The walk culminated in the ancient landscape of Avebury, having walked at least 10 miles a day, sometimes more, in circuitous routes to get there, and camping in eccentric places (with prearranged permission from the landowners). By then we had entered a state of deep landscape immersion, sensitive to the native plants and creatures, alive to the changes of each hill and valley, and with a feeling of wellbeing that is hard to explain. Walking, or tramping, had become our meditation.

Our bodies sunk into an awareness that was deeply peaceful. Phillip O'Connor, an articulate vagrant, describes this as an "incomparable feeling ... as though one were a prayer winding along a road ..." He found that during long periods of tramping a deep mental rhythm, 'poetic in effects', began to dominate all his perceptions. "All hard nodules of concepts are softly coaxed into disbursing their cherished contents ... One's 'identify-sense' becomes 'diffused into the landscape'..."† Linear time became irrelevant. We felt immersed in deep time, long, slow and luxuriantly smooth. Avebury itself came alive to me as Graham led us through the ancient turning of the seasons symbolically, the wheel of the year our ancestors so revered that they built great monuments to mark and celebrate this circular time and the miracle of birth, life and death.

I learned so many things in that week: to love the chalk Downland landscape even more deeply and to understand it through protracted observation; to appreciate the meditative quality of walking and the aliveness it brings to the body; to relish the warmth and humour of community, how gentle and generous people can be and how we all yearn for this intimacy in our lives; and to feel a connection to our ancient past. All this with consuming little and without travelling far...

The turning of western culture from destructive over-consumption to living within our limits can seem impossible whilst we are trapped in the hubbub of normal life, yet I saw in those few days how the future could be and it demonstrated to me the power of living simply in a group and being utterly alive to the elements. I am reminded of the Aboriginal rite of passage, Walkabout, and the tradition of indigenous people to slough off technological time and return to 'deep' time. This is a cyclical rather than

By then we had entered a state of deep landscape immersion, sensitive to the native plants and creatures, alive to the changes of each hill and valley, and with a feeling of wellbeing that is hard to explain.

linear consciousness that honours the seasons, the ancestors and the process of birth, life and death. These events were celebrated by my ancestors in the extraordinary Neolithic structures found at Avebury, West Kennet Long Barrow, Silbury Hill, the great Avebury avenues and circle of sarsen stones and the simple spring at Swallowhead.

If you only follow one of these suggestions in this chapter I would suggest you find the ancient trackways of your own country and walk in the footsteps of your ancestors. Find their campsites, explore their ancient settlements, discover their sacred sites and immerse yourself in your own bioregional landscape. We all hold an awareness of the indigenous within us, of a past rich with symbolism and wonder for the natural world. Our reconnection with this sense of reverence for Creation is a deeply joyful practice and will enable you to become a passionate advocate of our living planet, Mother Earth. And if you listen very carefully, She may sing Her sweet music to you once again, the song of the Earth.

† *Symbolic Landscape*, Paul Devereux, Gothic Image Publications, 1992, p38-39

Please see page vi for Maddy Harland's biography

Stephan Harding offers us a simple visualisation that allows us to appreciate both the grandeur of our universe and the wonder of our living 'animate' Earth.

Sensing The Round Planet

Stephan Harding

Lie Down on your back on the ground in your Gaia place. Relax and take a few deep breaths. Now feel the weight of your body on the Earth as the force of gravity holds you down.

Experience gravity as the love that the Earth feels for the very matter that makes up your body, a love that holds you safe and prevents you from floating off into outer space.

Open your eyes and look out into the vast depths of the universe whilst you sense the great bulk of our mother planet at your back. Feel her clasping you to her huge body as she dangles you upside down over the vast cosmos that stretches out below you.

What does it feel like to be held upside down in this way – to feel the depths of space beyond you and the firm, almost glue-like support of the Earth behind you?

Now sense how the Earth curves away beneath your back in all directions. Feel her great continents, her mountain ranges, her oceans, her domains of ice and snow at the poles and her great cloaks of vegetation stretching out from where you are in the great round immensity of her unbelievably diverse body.

Sense her whirling air and her tumbling clouds spinning around her dappled surface.

Breathe in the living immensity of our animate Earth.

… sense how the Earth curves away beneath your back in all directions. Feel her great continents, her mountain ranges, her oceans…

When you are ready, get up, breathe deeply, profoundly aware now of the living quality of our planet home.

From Animate Earth: Science, Intuition & Gaia, *Stephan Harding, published by Green Books in the UK by Chelsea Green in the USA, 2006.*

Dr. Stephan Harding oversees the MSc in Holistic Science at Schumacher College in Totnes, England. With degrees in Zoology and behavioural ecology, he has done field work in Zimbabwe, Peru, and Venezuela, and was Visiting Professor at the National University in Costa Rica before returning to England to help found Schumacher College in 1990. Stephan is the author of *Animate Earth: Science, Intuition and Gaia* (2006).

HAIKU originated in Japan among nature lovers and spiritual wanderers. It is based on pure economy of form and extreme generosity of spirit. With a stroke of a pen and a liberated spirit, good HAIKU reaches high and deep, connects experience and conscious mind. It is the sound of a solitary flute in a vast symphony of life.

Japanese Haiku

MARTI

Samurai Journey

Walk tall,
speak little,

bend like bamboo.

Iya Valley, Shikoku

Zen Pupil

A plank of rough wood,

waiting to be carved
into endless forest.

Mount Hiei

Jazz in the Lotus Matrix

Padma-
garbha-
loka-
dhatu.

Todai-ji Temple, Nara

Pilgrim's Path

Footsteps fly between earth
and sky,

only hills and valleys
stand in the way.

Kirihata-ji Temple, Shikoku

Zen Master

Calm clear eyes
bend boughs,
catch rain,

a hundred moons in a hundred
drops of water.

Sojiji Temple, Yokohama

Mathematics

Black river
plus
silver moon

equals
crystal water.

Kyoto

Source: Japan Haiku by MARTI, Collection Nature and Evolution, *Auroville Press Publishers, Tamil Nadu, India, 2005.*

MODULE 4
Health & Healing

Contents

The World as a Holowave: Theory of Global Healing

Planetary Healing: A New Narrative

Healing Ourselves

The Power of Reconciliation and Forgiveness

The Intelligent Heart

Peace Circle Dialogues: I Am Because You Are

The Cracked Mirror

Maher – Rising to New Life: Interview with Lucy Kurien

Health in the Global South

A Healthy Lifestyle

The Dream of the Children

All disease is the result of inhibited soul life. This is true of all forms in all kingdoms. The art of the healer consists in releasing the soul so that its life can flow through the aggregate of organisms which constitute any particular form.

Alice Bailey, Esoteric Healing

Dieter Duhm offers a vivid perspective of a holographic world filled with interpenetrating energies and dynamic information. Duhm points the way beyond today's society that is 'high-tech in war and Neanderthal in love' to a new society of peace and harmony.

The World as a Holowave: Theory of Global Healing

Dieter Duhm

The universe is a living unit. All things contained in the world find themselves in a mutual coherent evolution. Anything that occurs spreads itself as a holowave through space and time. What occurs at one place is happening everywhere (in corresponding translation and on differing scales). No pebble can be moved without something happening to the whole. And no star can burn out, without something changing on earth. Every event of war and every event of love leaves behind its traces in the finer tissue of the biosphere. The world matter that is being nudged somewhere, vibrates everywhere. The world reacts to each and every event, from the molecular structures to the galaxies and from a delicate thought to the most complex systems of information. What we are doing right now can alter the molecular structure of our teeth, or the speed of the growth of our hair, or the behaviour of our dogs, or the hope of some people in the peace villages of Colombia.

Each thought, each action derives from a current in the world matter. We cannot undertake anything and cannot make new plans if we were not seized by a holowave of the world matter. Everything that manifests itself anywhere on earth comes out of the invisible waves of the great whole. We are not only receivers of world matter, but also its activists and its steering organs. With each action, with each thought something is being changed in the world matter. If only we were successful in building up complex information for peace at any one place in the world – that also has the ability to withstand counteracting forces – then this information can be effective in the entire noosphere of our planet. From such thoughts the project of the Healing Biotope emerged. The more such places (acupuncture points) will be developed, the tighter the ring for global peace will become.

The New Picture

The world is an oscillating continuum. What shows itself as massacres and wars are global oscillations which may show up in other places in a much more subtle manner. The global life on earth, including all psychic occurrences, is a continuum of frequencies. Nobody is uninvolved. What transpires in protected affluent families in the form of disappointed love, secret lies, and latent aggression mounts up in other places in the form of hatred and genocide. Newly created peace and reconciliation sends vibrations into the world circle and arrives everywhere.

We no longer live in an era of matter, but in an era of frequency and energy. Spiritual and psychic vibrations are of a more penetrating nature than material ones. Considering these points of view, a totally new picture for the possibility of the healing of our planet opens up. It amounts to entering the correct frequencies in the whole. According to the quantum physicist David Bohm, ten to a hundred people, whose vibrations match, can change the world.

We no longer live in an era of matter, but in an era of frequency and energy.

An organism is attacked by an illness that shows up in many sick places. The intelligent doctor does not treat each place, instead he treats the whole organism. Translate this idea into the global situation of our time and an intelligent concept no longer provides peace work to special places of crisis. Instead it concentrates on the healing of the global organism, of both humanity and the earth. It is not only that thousands of individual areas are ill. The whole needs an injection and a new energy. The noosphere needs new information. The nuclei of the cells need a new impetus to be able to send fresh steering impulses to the organisms.

Five Key Insights

The entire earth and the entire humanity form an integrated organism, a holistic system, which reacts as a whole to the healing impulses that are injected.

The key information of a new, non-violent code of life is directed to the inner sphere of community, truth, trust, love, Eros, and religion. The bifurcations that set the course for human evolution on earth by taking new directions – once the material prerequisites are fulfilled – lie in the inner realm of the human being. Peace work is healing work. Central to it is the building up of trust. If we ask what does peace mean, it is this: peace is trust. Trust between human beings; trust between lovers; trust between parents and children, young people and adults; trust between women who love the same man; trust between men who love the same woman; trust between nations; trust between human being and animal; human being and nature; human being and world. The deepest meaning lies in finding the code of trust. Healing Biotopes are places where conditions that can support the emergence and growth of perpetual trust are consciously created.

Neither moral appeals, individual conversions, nor individual spiritual exercise will produce the necessary change. Instead new communitarian and societal structures have to be developed to promote a continuous growth of truth, trust and solidarity.

Communities of the future will build up new structures of the inner sphere of human beings, and thereby produce a global field of transformative effects. A new code of life will emerge out of the functioning of holistic systems, and change planetary life.

Steering of Power

Human society is a steering system of cosmic and human energies. The steering of the available energy and whether this leads to fortune or misfortune depends largely on the mental, social and technological steering systems that our culture has developed. Streams of energy can be led to collide with each other to produce paralysis, fear, enmity, violence and psychosomatic illnesses (sexual urge against morality, obeisance against rebellion). On the other hand, the streams of energy can be steered in such a way as to harmonise with each other, thereby producing healing, love and cooperation on a higher and vaster plane. The world needs new steering systems for the guidance and healing of the energies contained in the world matter. New steering systems mean: new thinking systems, new concepts for life, and new systems for the cooperation with the developing forces of life.

Community

The new levels of order exist in new forms of community and in the cooperation with the fellow beings in nature. Higher forms of union are higher forms of consciousness.

The new levels of order exist in new forms of community and in the cooperation with the fellow beings in nature. Higher forms of union are higher forms of consciousness. In my opinion, it is not possible to find the key information and harness the core forces for an emerging world unless we ground all research in the development and building up of real life communities. Thirty years of community work – such as the Project of the Healing Biotopes in the ZEGG and Tamera ecovillages – reveal perhaps more than anything else, the depth of transformation needed for a paradigm shift to occur in the human sphere. The new code of life is a result of functioning communities, communities in which the usual conflicts of authority, recognition, money, sex and love have finally rescinded in favour of a humane structure of truth, transparency and mutual support. Especially in the areas of sex and love the course has to be set in a new way for truth and trust to be able to develop. The code of trust of the new communities is based on truth and transparency. For this to become possible, the erotic forces have to be fully accepted and integrated in the community life. The old separations of official sexual moral and secret desires have to disappear in favour of a sexual integration that no longer calls for either suppression or lies.

At this point, the old thinking has to be revised and a new ethical reflection has to take place that surpasses by far all known humanitarian manifestos. The development of non-violent future communities may be the deepest adventure of our time. The renowned biologist Lynn Margulis adds to this by saying: "If we want to survive the social and ecological crisis we have created then we might be forced to enter into totally new, dramatic community ventures."

The global concept of healing starts with the concrete development of functioning communities. This has obviously proven to be more difficult than any development of ion arms and genetic engineering. Instead people have been waiting for the Messiahs to come to release them. There exists a communitarian intelligence capable of developing the code of healing.

Code of Healing

The new code consists of new information – information that is contained in the human genes. These have to be newly combined and activated. Old chains of information of war, jealousy, fear and violence that have been tapped into for years and have thus been active need to be dissolved and replaced by new ones. The totality of life is built on information. People's fortune or misfortune and that of all fellow creatures depend on the information that is being activated. Nothing is fixed forever, almost every matrix can be realised if the appropriate information is activated. We live in a true multiversity with an endless number of latent 'parallel universes'. They can also be called 'parallel holograms'. There are holograms of fear and holograms of trust, holograms of estrangement and holograms of connectedness. Which one of the many possibilities will manifest itself in our lives depends on the information we draw from the world, as well as on the information we inject back into the world. This process might be understood more easily if we consider how it functions in a love relationship.

David Bohm spoke of the 'implicate order' that generates all outer appearances in the manifest world. What is being manifested out of the implicate order depends on the type of information, thoughts and visions that are being activated. Whether wishes are fulfilled or not, whether humane goals can be reached or not, whether we experience happiness or unhappiness in love, depends entirely on the chains of information which are being activated through our thoughts and behaviour. Whether we are to find a peaceful re-connection with the whole depends on the successful dissolution of the dogmas of patriarchal religion, which are hostile towards love, and replacing them by the kinder information of a 'Marian' era.

Information is no abstract thing; rather it consists of lively, mental impulses. Authentic information always has the power to realise something new. Albert Einstein said: "What can be thought can also be done." These are authentic thoughts and visions that move the latent content of the implicate order into an 'animated state' and direct it towards manifestation. Vision and reality are ontologically connected to each other.

Fields of Information

We live in a holographic world. All information of the world is available and can be retrieved at any one point. Theoretically, the overall world information can be 'downloaded' at any place, at any given point and with the needed frequency. It is possible to tap into the world information or certain fields of it, like for instance into the field of mathematics or of music. We all know the phenomena of so-called child prodigies. These children command musical or mathematical capabilities by far surpassing our imagination. Although these children mostly do not have above average intelligence, they seem to have some sort of antenna with which they connect to parts of the universal database in a special way. Our community once had a guest staying with us who was able to continuously produce Goethe-type poems without ever having studied Goethe. For some reason he was connected to a 'Goethe field'. There is a man living in England with a low IQ, who, when sitting at a piano is able to play 50 pieces of classical music free of any mistakes. He is obviously connected to a field of music. People such as Houdini, clairvoyants and other mediums, some mental healers, yogis, circus acrobats, people doing extreme sports – all these have PSI ('psychic') capabilities. PSI does not conform to the usual notions of our physical world. These people are connected to information beyond those of conventional physics.

World information, whether known or unknown, is everywhere. At this very moment and in this very place where you are sitting and reading this, there are radio waves of all the broadcasting stations worldwide, as well as the radio waves of all the galaxies of the universe. I only have to switch on the radio and go to a certain wavelength in order to call in information. Similarly with the TV and the internet: the invisible presence in space can become something visible,, for instance in the form of a picture or written words. Marvel over marvel! The invisible world is full of information and frequencies. Were we without receivers, we would know nothing about it! The same goes for the flammability of wood, if we did not have a match to ignite it, we would not know about it. (How enormous is the difference between the visible world and the real possibilities she harbours; let us now transfer this to people!) An enormous invisible world full of possibilities is always present.

Frequencies always contain information. Which frequencies we take up depends entirely on our sensory capabilities, our antennas, and our intellectual and spiritual orientation. We are capable of refocusing our antennas and reinforcing the healing frequencies through brainwork, meditation and prayer, through music and dance, through suitable nutrition and physical training. In principle, there is not one single possibility that remains closed to us. Even in the darkest cellar healing can happen, as was shown by the French resistance fighter, Lusseyran, during his internment at the concentration camp of Buchenwald (see his book *And It Was Light Again*). At each place on earth the 'sacred matrix' is active, that is, if our sensory functions are tuned in and aware of its presence. All information necessary for the healing of human beings exist on earth and are available for retrieval if we are successful in building up suitable base stations. These base stations are the human organism and the human community.

We are capable of refocusing our antennas and reinforcing the healing frequencies through brainwork, meditation and prayer, through music and dance, through suitable nutrition and physical training.

Field Building

How and why can the new healing information be spread globally? The answer is to be found in the special functioning of holistic systems. Information compatible with the system and entered into it affects all its parts (see the above analogy about sick organisms). All parts are connected with each other by frequencies and are in resonance. Humanity, together with the whole of the biosphere, is a holistic system. What I do unto animals has an effect on people and what I do unto human beings has an effect on animals. The basic life structure is the same in all beings and the information structure of the genetic code is also part of this. It is to be found in all beings on earth – in the stinging nettle, as well as in the earthworm, the shark and the human being – principally it is everywhere the same. The differentiation is the way each being expresses itself.

All beings are part of 'one being' and 'one consciousness'. Frequencies, information and energies are all connected with each other in continuous cycle – all are part of the whole. Together they build a biological internet with running radio contact between all participants. This, according to Sheldrake, creates such phenomena like that of 'morphogenetic fields,' which respectively build 'morphic resonance'. If a new impulse is entered into a holistic system, then it affects all its parts. If the impulse contains new core information then a field-like change occurs, which makes itself noticed either as a 'mutation', an 'evolutionary leap' or as 'transformation'. Such transformations occur in the lives of individuals as well as in the lives of entire populations. All of evolution seems to have taken place through the impact of such field-like mutations. If we are successful in effecting a change in the nucleus in the areas of love, trust, sexuality and community, then this event will undoubtedly call forth a marked change of direction in human evolution.

Such an inner event could have unforeseeable effects on outer situations. A society with human fulfilment in love and with functioning communities would no longer have to rely on vicarious satisfactions such as consumption, possession, power and boasting. The destruction of nature that has been connected to the compensatory needs of unsatisfied people would come to an end. The development of functioning, planetary communities in which the elementary needs of belonging and connectedness are fulfilled, would therefore have a fundamental ecological effect. The healing thoughts in the ecological, social, sexual and spiritual realms belong intimately together.

The iteration of new systems happens by 'infection' and resonance. Everywhere on earth we see the phenomenon of 'infection'. For that we need neither bacteria nor other material vectors. If one of two similar systems is changed, the change will probably devolve to the other one as well, the systems 'infect' each other.

This phenomenon is well known in physics and chemistry; we know it even better amongst people. A simple example: If one starts to cough during a lecture, many others start to cough right away. A latent cough field

The development of functioning, planetary communities in which the elementary needs of belonging and connectedness are fulfilled, would therefore have a fundamental ecological effect.

creates the chain reaction. A bigger example: If the youth of the western world come into an existential crisis and a philosopher like Sartre or Camus express this crisis, hundreds of thousands of young people in Germany and France read the literature of existentialism. When a new field of thought comes up in physics, for example the quantum theory, within a short period of time quantum theory develops (Congress in Copenhagen 1927) and it does not take long until it is accepted by all physicists. When the first graffiti artists leave their wonderful figures on the walls, we will soon see the same figures in all big cities, underground stations and bridge pillars of the world from New York to Tokyo. Keep these examples in mind! Why do the same figures suddenly appear in the same perfection on the walls of ruins in the Portuguese Alentejo, miles from everywhere?

If a latent field for a new action, a new development or a change exists in a population, then this latent field can change into visible reality as soon as the first elements effect it. As soon as the first cells of humankind are able to develop new systems for a free life and to build up functioning communities, the field will spread on earth and convert into visible reality at many places.

We live in an actual time of transformation; it is, to use Ernst Bloch's words, full of 'utopian latency'. New possibilities, existing in the information chains of the becoming universe, yearn to be realised.

Humankind has waged war for thousands of years and is still galvanized by banners and guns. The liturgies of violence repeat themselves again and again, from Homer to Hollywood. The same patterns, the same chains of information in the genetic code, the same synapses in the brain, the same hormones are activated again and again, the same paroles shouted and passed on to the next generations. They are wrong basic axioms of thought, which – never scrutinised – have led to the globalisation of cruelty; masculine axioms of law and order, of sexual suppression, sinful bodies and a punishing God. The one-sided selection of information in favour of power, war and repression has led to astonishing achievements in the technical area and at the same time to a backslide of the mental, ethical and emotional developments to primitive, pre-cultural structures. 'High-tech in war and Neanderthal in love': this is the cultural picture of today's civilisation.

Let us imagine where humankind could be today if it had used the intelligence at its disposal for love instead of war! A civilisation that produced self-navigating weapons and brings electronic labs to Mars would also be able to create completely new social systems where violence and lies no longer have an evolutionary advantage and where love is no longer linked to jealousy. It is not human beings' nature that is destroying our planet but the one-sided choice of our possibilities. We can no longer react with accusation and anger. The earth does not need our affects but our intelligence. We need a new base of culture and a new code of life on this planet.

Extracted from *Future Without War: Theory of Global Healing* by Dieter Duhm BoD, 2007.

We live in an actual time of transformation; it is... full of 'utopian latency'. New possibilities, existing in the information chains of the becoming universe, yearn to be realised.

Dr. Dieter Duhm is co-founder of the Tamera community in Portugal, and author of several books including *The Sacred Matrix* and *Future Without War*. Trained as a psychoanalyst and sociologist, he left his earlier academic and political work to found innovative communities for 'Peace Research' and 'Healing Biotopes'. Today he is creating a centre for art and healing, and is working to build a global cooperative for a future without war.

Maddy Harland explores an energectic appreciation of landscape and ways in which we can combine restoring ecosystems with practical and scientific knowledge, and with a deep connection with the Earth as an animate, intelligent being.

Planetary Healing: A New Narrative

Maddy Harland

The Landscape

New Gaian science tells us that our planet is 'animate', an intelligent, living, self-regulating being. For millennia, yogis have taught that our physical bodies are animated by elemental energies or intelligences that we can co-operate with in healing and that we are indeed a microcosm in a larger whole. The landscape that surrounds us is a 'body' and it has an integrity and intelligence that is governed by the laws of nature. It is possible to study these laws and work with these subtle energies.

There are a number of practices, ancient and modern, in the field of Esoteric Agriculture which are a step beyond purely organic growing systems. Many have heard of the biodynamic system of growing developed by anthroposophist, Rudolph Steiner. Biodynamics is a practice that works energetically as well as on a physical level and includes planetary influences on plants, such as the phases of the moon and the constellations. Many of the famous chateaux in France use biodynamic methods, though they may not always promote this fact on their labels, because these methods increase yields and help produce excellent wines.

In Australia, the fragile, ancient and brittle environment is easily exhausted and damaged by European agricultural methods. Alien plant species like bramble can run riot, as do alien species such as the rabbit. Over grazing and intensive arable systems quickly deplete the top soil. Consequently, farmers are learning how to work with the subtle energies of the Earth to heal it and enhance fertility. Naturally, some of these practises are ancient in origin, well known to our ancestors. Briefly, there are a number of them.

Lithopuncture

Lithos – stone (Greek), *Pungere* – to pierce (Latin)

The idea of Lithopuncture was developed by Marko Pogacnik from Slovenia though it has ancient origins. He identified that human developments such as inappropriately placed buildings, dams and roads can cut across ley lines, the earth's meridians, and disrupt energy centres that are akin to chakras in the landscape. He worked redirecting or rerouting them and restoring balance initially by driving metal rods into the ground like an acupuncturist at specific points. He later evolved this technique by discovering that he could place appropriately sited stones at the same points to restore the flow of energies.

In Austria, the highways authority has started using these monoliths to prevent accidents at black spots. On the A9 Pyhrn motorway in Styria the authorities erected two massive white quartz megaliths with the help of a Druid geomancer, Gerald Knobloch. They were located away from the road and the project was kept secret. Previous attempts at warning signs had failed to eradicate the problem. Over a two year period the accident rate fell from six fatalities per year to none. The 100% success rate cannot be disputed and the Austrians intend to use this technology in blackspots elsewhere. Knobloch explains that rerouted water flows, bridges and construction projects disrupt natural energy flows which affect drivers and create 'sleep zones'.[1]

Australian esoteric agriculturists, with the help of the dowsing community, have also discovered other beneficial effects from placing stone megaliths in gardens and on farms – fertility and plant growth was enhanced (as well as human health). This practise has been refined by a permaculture practitioner, Alanna Moore, who teaches people how to build 'power towers', based on an ancient Irish idea. Basically, she sinks a hollow tube designed with specific dimensions into the ground, fills it with paramagnetic rock dust and constructs a special cone to cap it. The location of this devise is dowsed for maximum effect. There are apparently hundreds being tested around Australia by commercial organic and biodynamic farmers as well as gardeners and anecdotal evidence of increased yields and better plant health.

On the A9 Pyhrn motorway in Styria the authorities erected two massive white quartz megaliths with the help of a Druid geomancer, Gerald Knobloch... Over a two year period the accident rate fell from six fatalities per year to none.

In The Community

Acupuncturists use energetic pathways (meridians) in the human body and locate 'points' to increase the flow of healing energies along those pathways. Similarly, esoteric agriculture recognises flows of energy in the landscape. In Europe, there is also a thousand year old tradition upheld by the Master Masons. Our great cathedrals were built by craftsmen who were working with subtle energies in the form of architecture. The design of the cathedral symbolises the crucifixion form – the intersection of spirit (vertical) with matter (horizontal). It embraces the universe, according to the author, Peter

Dawkins, in love, in joy, in wonder.

These craftsmen had an understanding of the chakra system of the body and applied it to building. The nave, where most people sit, is located in the Solar Plexus chakra – the building's abdomen. Where the transept crosses the nave (called the crossing) is a central tower rising up, marking the heart centre. The heart is a vital part of the building. The chancel has the chakras associated with the head; the throat where the Word is spoken and sung; the Ajna where the bishop sits (called the cathedra) where he can view the cathedral and co-ordinate the proceedings; and then right at the east end, the crown chakra where the high altar is set, the place of final wonder, joy and communion. The font is usually placed in the root chakra, the ceremonial

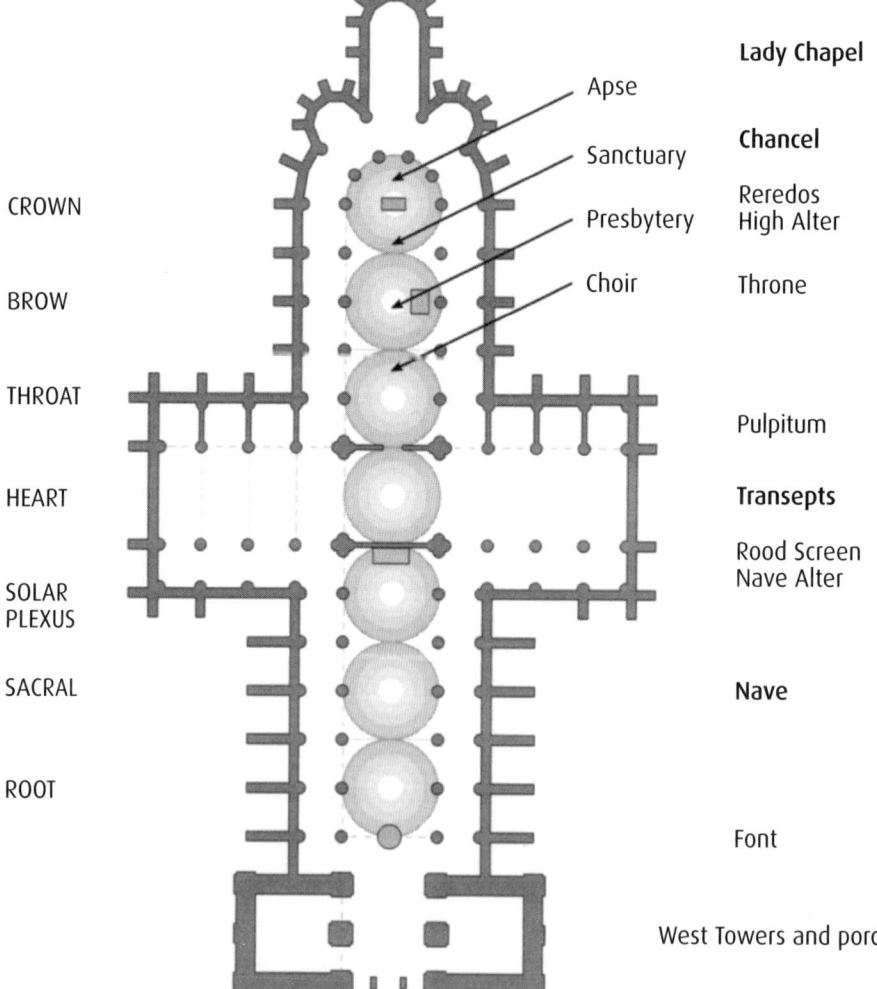

Plan of Cathedral.
© Peter Dawkins

place of entry into this world and of new beginnings.

Imagine constructing communities with this knowledge of chakras and their energetic connections. Peter uses Findhorn as an example of what is possible. The foundation of the community, the original caravan where Peter and Eileen Caddy lived with Dorothy Maclean at the very beginning, is found at the heart centre of that community. Food is appropriately served in the solar plexus, the Community Centre, and gatherings, conferences and concerts are held in the head centre, the Universal Hall. The bookshop and businesses are located in the sacral area and the root chakra is home to the living machine – an ecological waste processing plant that recycles the water and sewage for the community. This placement has evolved intuitively in the community but there is no reason why we cannot employ the knowledge used by the master masons for future settlements.

Equally, as community is a 'body', we can work energetically with it. If we understand the maxim that 'Energy Follows Thought', by using prayer, evocation and visualisation we can encourage, heal and connect with the 'genus loci', the spirit of a place. It follows then that the people within the community will also feel the benefit.

Our planet is an exquisitely beautiful being, alive and intricate, just like each one of us, and evolving as it revolves in its own cycles.

Planetary Healing

There are many great forces at work within our planetary system that we barely understand. This does not prevent us from lending our co-operation. There is a growing movement for planetary healing that in part has stemmed from the more esoteric aspects of the environmental movement. There is also growing spiritual movement seeking planetary harmony and these two are beginning to converge.

There is an incredible amount of nonsense spoken and written about planetary healing in the new age movement. It is important to keep our focus on goodwill and the tenor of our thoughts rather than focus on the cosmic glamours that abound. We cannot ever fully understand the great process of our planet's evolution and nor should we underestimate the power of the Earth's capacity for regeneration, with or perhaps ultimately without the presence of humanity on this planet (if we do not curb our capacity to expel greenhouse gases into the atmosphere, rendering our home inhospitable to human beings).

In terms of healing, it is useful to keep things grounded and simple. We can use the power of thought to link together with all people of goodwill and imagine a powerful web of light surrounding, nourishing and healing our planet and the consciousness of its inhabitants being raised. We can work with the cycles of the Moon to deepen our spiritual understandings and insights. This practise regards the time leading up to the Full Moon as a time of greater opportunity, a time when more light can be cast for greater awareness.[2]

We can meditate for greater peace, understanding and love between people, communities and nations. We can also send out our love and appreciation to this exquisite home that we inhabit. I believe our planet is an exquisitely beautiful being, alive and intricate, just like each one of us, and evolving as it revolves in its own cycles. This being deserves our appreciation and respect and all our efforts towards peace, harmony, and the will-to-good. Our meditation practise has much potential scope to develop this.

Two Edges Meet – Combining Spirit with Matter

There are now people all over the planet who have the technical capacity to restore damaged ecosystems. They combine scientific and practical knowledge with a deep, enduring commitment to the wellbeing of the Earth and its species, including us, and they hold insights into nature's intelligence. This combination of scientific disciples such as environmental science and ecology with an ethos of service to humanity and the planet is powerful and effective. Below are just three examples but I could fill an entire book with stories.

By piecing together a complex ecological puzzle, biologist Willie Smits has found a way to re-grow clearcut rainforest in Borneo, saving local orangutans – and creating a thrilling blueprint for restoring fragile ecosystems.

By piecing together a complex ecological puzzle, biologist Willie Smits has found a way to re-grow clearcut rainforest in Borneo, saving local orangutans – and creating a thrilling blueprint for restoring fragile ecosystems. Dr Chris Reij, a specialist in sustainable land management from the Netherlands, who has been developing techniques to successfully rehabilitate degraded land in the Sahel, a vast strip of land across the southern Sahara, running from the Atlantic Ocean to the Red Sea. Already, 50,000 square kilometres (21,000 square miles) have been greened. Finally, Austrian farmer and permaculturist, Sepp Holzer, has transformed his own spruce monoculture in the Austrian Alps 1,500 meters above sea level. His farm is an intricate network of terraces, raised beds, ponds, waterways and tracks, well covered with productive fruit trees and other vegetation, with the farmhouse neatly nestling amongst them. Now Sepp works many countries of the world sharing his knowledge. I met him at Tamera ecovillage in Portugal and saw for myself the combination of practical experience, technical knowledge and a deep spiritual connection with the elements and the land can literally turn back the desert.

I had heard that Sepp Holzer was building lakes in the arid Iberian Peninsula in Portugal at Tamera ecovillage. Visiting in February, I imagined that they would be reasonably impressive after the winter rains. I had even watched a webcam of the first lake filling up in 2010 but I had no concept of the scale of the restoration work, however. Sepp and the Tamera community have literally dammed a valley and stopped the rain and topsoil rushing down the valley and into the sea.

Ever decreasing yields in this depopulated, rural region have driven poor farmers to try and extract more from the land than it is capable of bearing. Sheep are stocked at such density that the pasture is destroyed. Nature responds to the overgrazing by growing a 'scab', the inedible rock rose, *Cistus* spp., often the first pioneer after wildfire. 90% of the remnant cork oak forests are dying due to soil compaction that destroys soil mycorrhiza. The rest are being felled as cork falls out of favour, replaced by eucalyptus, hungry exotics in a brittle landscape. With the oaks dies a unique, biodiverse habitat and the Iberian lynx and Bonelli's eagle are threatened with extinction.

Sepp and the Tamerans have reversed this process in their valley. They stopped the overgrazing and ploughing, focusing the community's food production on fruit and vegetables. There are raised beds everywhere full of annual and perennial vegetables. Fruit and nut trees line the banks of the lakes. The winterbourne stream is dammed and there is an interconnected system of lakes that flow into each other as the slope falls down the valley. It is almost unbelievable that in such an arid landscape, so much water can be collected. This is living water too, with rippling surfaces, filled with frogs and fish, to keep the balance between mosquitoes and humans healthy. Sepp cups his hands and tells us, "God gives us enough water. All we have to do is find a way of holding it in the landscape."

What has been achieved in the first three years is astonishing. Early morning mists arise out of the lakes and leave their dew on the surrounding plants. Swallows swoop and drink. Otters have returned. New springs rise in the surrounding hillsides. The younger cork oaks are growing and seeding. Even a Bonelli's eagle has visited. Perhaps it will return with a mate. The whole landscape is being reaquified with living water.

The pulse of our animate, intelligent planet pulses through me. My heart opens with the knowledge that we can restore the Earth. I hear stories of wonderful projects and people rebuilding soils and regenerating exhausted lands, cleansing waterways, reforesting bioregions, developing new regenerative agricultural practices, and applying innovative ideas the world over. They all share a reverence for our beautiful planet, insights into how to work with the laws of nature and her inherent intelligence. The Earth itself responds and the capacity to repair and regenerate ecosystems is far faster than we can imagine. These stories that reverse our sense of despair and disintegration must be a part of a new narrative.

Sources

1   *The Sunday Times*, June 1 2003.
2   *Reflections on the Rhythms of the Year*, Michal Eastcott, Sundail House.

Healing The Heart of The Earth: Restoring the Subtle Levels of Life, Marko Pogacnik, Findhorn Press, 1 899171 57 6.
Stone Age Farming: Eco-Agriculture for the 21st Century, Alanna Moore, Python Press, 0 646 41188 8.
Creating Harmony: Conflict Resolution in Community, ed. Hildur Jackson, Permanent Publications, 1 85623 014 7.
Sepp Holzer's Permaculture, Sepp Holzer, Permanent Publications, 978 1 85623 059 9.

Please see page vi for Maddy Harland's biography.

Our medicine reflects our cultures. Both allopathic and complementary practices have their place. Maddy Harland suggests that healing takes a step further beyond medicine and is a mechanism for the evolution of consciousness.

Healing Ourselves

Maddy Harland

The Pachamama Alliance in both the North and the South speak of inspiring a 'New Dream', a vision of the world to yet to come. They speak of the bird having two wings; one wing being the global South with all the intelligence and wisdom of the indigenous peoples, and the other the technical, analytical and innovative capacity of the people of the global industrial North.

Medicine is also a bird with two wings, one wing the traditions of holistic medicine, many ancient, and the other the new scientific medicine of the industrial North. Our modern allopathic medicine sees the body as the sum of its parts and therefore sets out to treat symptoms when it malfunctions. This is not to say that the parts are not magnificent. The blood and lymph systems are a marvellous architecture throughout the body. The rhythmic way in which the heart beats, stimulated by electrical impulses, is poetry whilst the potential of 90 per cent of our intricate brains still remains a mystery to us.

We are prescribed drugs and sometimes placed under the surgeon's knife to remove an offending organ or tumour, fix a fracture or repair a wound. This is science at its most practical and useful. If I broke my wrist I would be deeply grateful for the skills of a surgeon and if I contracted pneumonia I would bless the invention of penicillin. My grandfather was not so fortunate and died in his 50s in 1936 from pneumonia. There is no doubt many of our drugs are invaluable *in extremis*. Yet just as the global North has industrialised and chemicalised our agriculture, we have also allowed our medicine to become commercialized by pharmaceutical companies and mechanized by reductionist science. We often turn to drugs and surgery by default. We have lost respect for our heritage of folk medicine and we have rejected any practice that cannot be 'proven' by the narrow confines of materialistic, reductionist science.

The wisdom and insights of the herbalist, shaman, and yogi has been sidelined in favour of conventional approaches. We have become patients waiting for an expert to cure our symptoms, divorced from a sense of our own responsibilities and evolving consciousness, passively waiting to be fixed. Yet even with the entire healthcare advances of modern science industrialised societies face a crisis in wellbeing. We are fatter and more riddled with heart disease, diabetes, clinical depression and cancers than at any other time in human history. Whilst still appreciating the great privilege of industrialised medicine in a crisis, we need to evolve into a more responsible way of being healthy and treating dis-ease.

There is no doubt that 'energy' medicine and medical herbalism work. Babies respond well to homeopathic, immune to the persuasive argument of the placebo effect. Women can give birth relatively pain free with the help of a skilled acupuncturist. It is easy to make simple tinctures and herbal preparations to ameliorate acute symptoms like coughs and colds and sore throats with common herbs that we can grow in our gardens. With many herbal medicines being banned in Europe, there is a growing movement to return to our folk roots and once again grow and harvest our own medicines at home.

How Does Energy Medicine Work?

Lets take Acupuncture as an example. Acupuncturists understand that the body is not just a bundle of organized electrical impulses and a collection of organs and systems. It is a living energy field that has channels called meridians and the life energy itself is called *Qi*. Qi flows through the meridians that cross and create a web of points that can be stimulated. By placing tiny silver needles below the surface of the skin we can stimulate these points and correct imbalances in the flow of Qi through the meridians.

Acupuncture is an ancient Chinese medicine and its origins are believed to date back to 1600-1100 BCE. Yet in most cultures there is a tradition of energy medicine and knowledge of herbalism that understands the body to be more than the sum of its parts but an interrelated energy system through which life flows. Disease results from the interruption of that flow and the practitioner's role is to encourage the release of the obstruction so that the energies can rebalance.

Is this so far fetched? Our world is alive with energy. Faraday and Maxwell have taught us that invisible electromagnetic waves travel through space. We have also discovered that our nerve impulses are electrical and the heart is not merely a muscle or pump but an electrical system. A *New Scientist* article 'Healthy Vibes' describes a small copper sensor has been devised by Terry Clark of the University of Sussex, and engineers Robert Prance and Christopher Harland. It can pick up the body's electromagnetic field from a metre or more away and was invented because contact with the skin drains and distorts the signal. This 'super sensor' produces the most sensitive electrocardiograms to date.

Our world is alive with energy. Faraday and Maxwell have taught us that invisible electromagnetic waves travel through space. We have also discovered that our nerve impulses are electrical and the heart is not merely a muscle or pump but an electrical system.

Energy and Matter are Interchangeable

William Keepin's first chapter in this book on science and spirituality builds on a fundamental understanding developed by Einstein: that energy and matter are interchangeable. Using sensitive detectors, it has been found that light – an electromagnetic wave – can also behave as if it is made up of particles (photons). This works both ways: sub-atomic particles have been found to have waves associated with them. Quantum theory has produced a theoretical acceptance that material substance has both a material form and a wave (energetic) form associated with it.

Blavatsky elucidated the idea that of matter being energy vibrating at its slowest rate in *The Secret Doctrine*. She speaks of the entire universe being "that infinite Ocean of Light, whose one pole is pure Spirit lost in the absoluteness of Non-Being, and the other, the matter in which it condenses, crystallizing into a more and more gross type as it descends into manifestation" (*The Secret Doctrine*, 1:481). Material particles, she said, were 'infinitely divisible centers of force', and matter could therefore exist in infinitely varying degrees of density. Our physical senses have been evolved to perceive only one particular plane of matter, which is interpenetrated by countless other worlds or planes invisible to us which is composed of ranges of energy-substance finer than our own. This was known to the ancient Rishis of the Himalayas, the Taoists of ancient China and to the indigenous peoples the world over.

All forms of energy healing acknowledge the existence of fields of energy in and around the body. We are more than the sum of our parts, whether they are physical organs and systems or electrical fields. Our bodies consist of a physical form – energy vibrating at its slowest rate – but also more subtle invisible forms. "Most complementary therapies work to assist in the rebalance of the 'life energy' of an individual, recognising that blocked or disturbed energy is often the cause of illness or negative symptoms. This energy flows through and around each of us in a very specific way, creating a subtle energy field that informs and gives vitality to our physical form. The energies generated by our emotional and mental states also affect this field," explains Dinah Lawson, a teacher from the International Network of Esoteric Healing.

Esoteric Healing

Esoteric philosophers have termed one of the less visible of our forms the etheric body. Alice Bailey, who wrote *Esoteric Healing*, says 'The etheric body is a body composed entirely of lines of force and of points where these lines of force (the meridians described by acupuncturists) cross each other and thus form (in crossing) centres of energy.' These lines of force are known as chakras. We have seven major chakras, 49 minor chakras and countless lesser ones throughout the body. We can learn to sense these forces of energy just as dowsers sense changing energies in the landscape and we can also visualize that the balancing of major and minor chakras.

Our physical senses have been evolved to perceive only one particular plane of matter, which is interpenetrated by countless other worlds or planes invisible to us...

This is in essence what subtle energy healing is – a way of sensing and working with the subtle energies of our being. These subtle energies, like Einstein's theory of field, extend far beyond the electro-magnetic field around our bodies. They energise our emotional, mental and spiritual lives. Our lives are a journey of understanding and our individual challenges are to overcome obstructions in our energy fields and to seek an expansion of consciousness in on every level. We are here to learn how to become more physically balanced and healthy, more emotionally intelligent and mentally refined and psychologically at ease in the world. Ultimately, however, our journey is a spiritual one; to realise the profound interconnection of all beings within the web of Life; to realise that we are One.

The study of Esoteric Healing provides useful tenets to guide all practitioners whatever the discipline:

> *An esoteric healing practitioner does not claim to heal, but rather to assist (where permitted) in the re-balancing of the energy field of an individual. This is in order that the person's own life energy can flow freely again and aid the healing process. The practitioner will align to the Source or Spirit of all Life so that energy will flow through him or her, and will offer – but never force – energy, and ask that healing shall be given in accordance with the will of the client's higher self or soul.*
>
> www.ineh.org

Our lives are a journey of understanding and our individual challenges are to overcome obstructions in our energy fields and to seek an expansion of consciousness in on every level.

The whole person is treated, rather than the disease, and this varies according to the needs of the individual. By connecting with the source of all Life and offering energy, the practitioner is not working 'on' someone, imposing personal their will, they are working with the person's blessings and with the flow of energy. This work is based on a practical understanding of the chakra system and how it transforms energy down through the nervous, blood and lymph system to the major endocrines and then the organs and bones. The underlying philosophy, described in detail by Alice Baliey in the book *Esoteric Healing*, is as ancient as yoga itself. Whilst travelling in Bhutan I visited the school of traditional Bhutanese medicine that was introduced to the country by the Tibetans in the 8th century. Many of the charts in the school were representations of this knowledge, also to be found yogic texts.

Birth

In addition to an understanding of healing as the rebalancing of energy flows is also the wider understanding that our lives are not merely a biological 'one off'. New-borns are not *tabula rasa*[1] but are souls reincarnating often many times over vast spans of time. Hildur Jackson says that we need to treat children with utmost respect and love during pregnancy, when they enter the world and in childhood and minimize medical interventions as much as possible. These ideas are important to the emerging ecovillage and cohousing culture which tend to create environments that are more able to uphold these values. Where there is an enlightened community, there is more support for the young family.

Death

In Western culture, we have lost our appreciation of both the role of elders and the abundantly creative role that youth can play in shaping society. We have also lost our connection to the great mystery of death. Instead we hide death away and try and sanitise it. Death has become an ending to avoid at all costs, a fearful and angonising process to be controlled by medical intervention. Yet there is a growing understanding that death is not a final ending but a transition that we can learn to approach with dignity and calm. If we are indeed beings of energy, the death of our physical form is like shrugging off a garment we no longer need. For all of our lives we can learn to practise 'letting go' – of the material world, our attachments, our fears, our passions – so that rather than our final days being a terrifying loss, they can be met with peaceful acquiescence. There are many accounts of yogis being able to leave their physical bodies at will. Some enlightened people have been reported to leave their bodies at will and at an appointed time. We may not achieve this level of control but we can 'die well'. Indeed, I have watched a number of people let go calmly once they felt their worldly duties were completed. Death can be a beautiful transition in these circumstances and indeed Esoteric Healing can assist the process of letting go. Furthermore, a deeper understanding of death is fundamental to the spiritual life, and to helping us define and realize our deepest priorities and purpose in life.

Many spiritual teachings propose that life itself is a preparation for learning to die well, learning to let go of all attachment and fears and moving on to a new phase of awareness. Esotericists call the life of an incarnation 'limitation' and the death of the body 'liberation'. When we realize that death is a transition to consciousness beyond the individual life, beyond the material form, we begin to view death in a positive rather than a negative way. These teachings tell us that the only real death is limitation – which is precisely what incarnation into a dense physical body is – from the perspective of the enlightened mind. Liberation from this limitation (the physical body) is thus actually an entrance into greater life. These teachings turn conventional thinking on its head. We are conditioned to believe that death is the end and life should be held on to at any cost. Yet here we see that what we call 'life' is actually a confining limitation, and what we call 'death' is actually an expansive liberation. With death as liberation from the limitation of form, we begin to understand why yogis could chose the time of their death and why dying can be a gentle transition. In *Esoteric Healing* by Alice Bailey, she writes:

> *Know that that light descends and concentrates itself; know that from its point of chosen focus, it lightens its own sphere; know too that light ascends and leaves in darkness that which it – in time and space – illumined. This descending and ascension men call life, existence, and decease; this We Who tread the Lighted Way call death, experience, and life.*

Life Between Lives and Reincarnation

Many cultures have reincarnation as an important part of their philosophy and belief. The Dalai Lama, in his autobiography, describes his last lives and how he was recognised as the 15th reincarnation in this life as a small boy and brought up in a monastery. Most cultures believe in reincarnation apart from the official Christian who declared it a formal heresy at the Second Council of Constantinople in AD 553. Prior to that time, it had been a fundamental Christian teaching. *The Phoenix Fire Mystery Book on Reincarnation* explores religions, philosophies, science, literature and demonstrates that many of the great sages of the world actually had an idea of who they have been in earlier lifetimes and where they had lived. This helped them to find the purpose of their present life.

Psychiatrist Ian Stevenson investigated many reports of young children who claimed to remember a past life. He conducted more than 2,500 case studies over a period of 40 years and published twelve books, concluding that "reincarnation is the best – even though not the only – explanation for the stronger cases we have investigated." Although his research is controversial, the conservative scientist Carl Sagan acknowledged the validity of Stevenson's data, and recommended further serious study.[2]

The Future

If we regard personal healing as our responsibility, then the benefits of western medicine – so reliant on oil to synthesise, package and transport its pharmaceuticals – should only be enjoyed when absolutely necessary, rather than consumed passively as the only pathway to health. I imagine a future when it is once again normal for people to grow medicinal and useful plants in biodiverse, organic and edible gardens, and to harness their therapeutic properties by making healing balms, teas and tinctures. As science becomes more sophisticated in its ability to quantify energy, I also envisage a more enlightened understanding and appreciation of the efficacy of energy medicines.

Beyond that, the understanding that human evolution is not simply the survival of the fittest but an unfolding collective process of expansion of consciousness will change how we live together in community and ease the age-old frictions between us. This expansion will lead to a reverence for and an ultimate connection with the flowing energies of Life itself. Though bodily, emotional and psychological wellbeing are important, this is the most profound healing.

> *We are all visitors to this time, this place. We are just passing through. Our purpose here is to observe, to learn, to grow, to love ... and then we return home.*
> Australian Aboriginal Proverb

Reference

Alice Bailey, *Esoteric Healing*, Lucis Press Ltd, 1972.

1. *Tabula rasa* is the theory that individuals are born as a 'blank slate' without built-in mental content and that their knowledge comes solely from experience and perception.
2. Carl Sagan, *The Demon-Haunted World* (1996).

Please see page vi for Maddy Harland's biography.

Duane Elgin outlines the stages of the reconciliation process, and gives inspiring examples that show how the power of love can reconcile injustice by moving humanity beyond reprisal to forgiveness and restitution.

The Power of Reconciliation and Forgiveness

Duane Elgin

The Power of Love

Compassionate love is a transformative power that we cannot quantify or measure, yet it brings incomparable strength and resilience into human relationships. "Love," said Teilhard de Chardin, "is the fundamental impulse of Life ... the one natural medium in which the rising course of evolution can proceed."[1] Without love, he said, "there is truly nothing ahead of us except the forbidding prospect of standardisation and enslavement – the doom of ants and termites."

A compassionate love can provide a vital 'social glue' to hold us together as we face the challenges ahead. If we pull apart, an evolutionary crash seems assured. If we come together authentically, however, we have the real potential to achieve an evolutionary bounce. And to pull together, we need to reconcile the many differences that now divide us. We need to discover harmony where there is now discord. We need to cultivate the respect and regard for others that ultimately come from a foundation of love.[2]

Love is the deepest connecting force in the universe, and thus a vital ingredient in our evolutionary journey toward wholeness. The unfolding of love is not different from the unfolding of awareness. Jack Kornfield, esteemed meditation teacher put it this way: "I will tell you a secret, what is really important ... true love is really the same as awareness. They are identical."[3] If we can learn the lesson that love will further our evolution and that the greater our love the greater our awareness, then we are aligned for success in our journey home. With love – or a maturing awareness – as a foundation, the hallmark of the emerging era could be the healing of our many fragmented relationships. If that were to occur, it truly is possible to

Love is the deepest connecting force in the universe...

imagine a future that works for everyone. With reconciliation, there is little doubt that an evolutionary bounce could happen.

A compassionate or loving consciousness has ancient roots, but it is taking on a new importance as our world becomes integrated ecologically, economically, and culturally. Because we now share one another's fate, it is increasingly clear that promoting the well-being of others directly promotes our own. We have reached the point where the Golden Rule is becoming essential to humanity's survival. This ancient ethic, which is found in all of the world's spiritual traditions, advises that the way to know how to treat others is to treat them as you would want to be treated. Here are some of the ways the Golden Rule has been expressed:

> *As you wish that men would do to you, do so to them.*
> Christianity (Luke 6 : 31)

> *No one of you is a believer until he desires for his brother that which he desires for himself.*
> Islam (Sunan)

> *Hurt not others in ways that you yourself would find hurtful.*
> Buddhism (Udanavarga)

> *Do naught unto others which would cause you pain if done to you.*
> Hinduism (Mahabharata 5 : 1517)

> *Do not unto others what you would not have them do unto you.*
> Confucianism (Analects 15 : 23)

As diverse and divisive as we are, the human family recognizes this common ethic of compassion at the core of life. This indicates to me that there is a basis for reconciliation within humanity.

Love and compassion not only have ancient roots; history also attests to their impact and enduring power. Compassionate teachers through the ages such as Jesus, Buddha, and Lao-tzu have all lacked wealth, armies, and political position. Yet as the late Harvard professor Pitirim Sorokin explains in his classic book *The Ways and Power of Love*, they were warriors of the heart, and have reoriented the thinking and behavior of billions of people, transformed cultures, and changed the course of history. "None of the greatest conquerors and revolutionary leaders can even remotely compete with these apostles of love in the magnitude and durability of the change brought about by their activities."[4] In contrast, most empires built rapidly through war and violence – such as those of Alexander the Great, Caesar, Genghis Khan, Napoleon, and Hitler – have crumbled within years or decades after their establishment.

The ruler Ashoka, who lived in India three hundred years before Jesus was born, is an example of the power of love in human affairs.[5] Prince

Ashoka was born into a great dynasty of warriors and inherited an empire that extended from central India to central Asia. Nine years into his reign, he launched a massive campaign to win the rest of the Indian subcontinent. Finally, after a fierce battle in which more than a hundred thousand soldiers were slain, the land was conquered. Ashoka walked the battlefield that day, looking at the dead and maimed bodies, and felt profound sorrow and regret for the slaughter and for the deportation of people he had conquered. He immediately ceased his military campaign, converted to Buddhism, and devoted the rest of his life to serving the happiness and welfare of all.

Ashoka's 37 years of benevolent rule left a legacy of concern, not only for human beings, but also for animals and plants as well. His decrees creating sanctuaries for wild animals and protecting certain species of trees may be the earliest example of environmental action by a government.[6] Ashoka's works of charity included planting shade trees and orchards along roads, building rest houses for travelers and watering sheds for animals, and giving money to the poor, aged, and helpless. The end of war and an emphasis on peace marked his political administration. All his political officers were encouraged to extend goodwill, sympathy, and love to their own people as well as to others. One of their main duties was to be peacemakers, building mutual goodwill among races, sects, and parties. His cultural activities promoted education and the arts of the stage, including the construction of amphitheaters. Sorokin sums up Ashoka's legacy as "a striking example of a peaceful, love-motivated, social, mental, moral and aesthetic reconstruction of an empire."[7]

Ashoka's compassionate rule established the largest kingdom in India until the arrival of the British more than 2,000 years later. The lion pillar, symbol of Ashoka, survives to this day as the official emblem of the Republic of India, and it is also found on every Indian coin and currency note. "Amidst the tens of thousands of names of monarchs that crowd the columns of history," wrote historian H. G. Wells, "the name of Ashoka shines, and shines almost alone, a star."[8]

Based on examples such as these, Sorokin concluded that love-inspired reconstructions of society that are carried out in peace are far more successful and yield much more lasting results than reconstructions inspired by hate and carried out with violence. Again and again, he found that "hate produces hate, physical force and war beget counterforce and counterwar, and that rarely, if ever, do these factors lead to peace and social well-being."[9]

Reconciliation does not mean forgetting the suffering and injustices of the past; rather, it means not letting the past stand in the way of opportunities for the future.

The Process of Reconciliation

Reconciliation does not mean forgetting the suffering and injustices of the past; rather, it means not letting the past stand in the way of opportunities for the future. When historic injustices are publicly acknowledged and realistic remedies are found, hurts from the past no longer stand in the way of collective progress. Freed from the need to continue blaming and feeling resentful, people can shift their focus from past grievances to mutual opportunities in the present and the future.

The process of reconciliation is complex and involves at least three steps: the injured need to be heard publicly, the wrongdoers need to apologize publicly, and if appropriate, they need to provide restitution or reparations.

Being heard is the first step in being healed. By listening to and acknowledging the stories of those who have suffered, we begin the process of healing. Our collective listening to the wounds of humanity's psyche and soul is vital to our collective healing.

In An Ethic for Enemies, his book on the politics of forgiveness, Donald Shriver Jr. explains that in popular usage, the phrase 'to forgive' is thought to mean 'to forget'. But, he says, that is not what forgiveness means: "Instead, 'remember and forgive' would be a more accurate slogan."[10] Forgiveness requires that we surrender revenge as a basis for justice. We need to call forth mercy and forbearance in order to break through the cycle of violence and counter-violence. Forgiveness also requires that the injured seek to understand the actions of the wrong-doer so as to restore the wrong-doer's humanity. As a final step in reconciliation, both parties need to create a new relationship so that they can live together in peace and mutual respect.

Archbishop Desmond Tutu knows more about the process of reconciliation than most of us do. He was the chairman of the Truth and Reconciliation Commission (TRC), which was established to investigate crimes committed during the apartheid era in South Africa from 1960 to 1994. He describes the logic of reconciliation in his country in this way. When apartheid ended, South Africa's black majority had to choose among three ways to seek justice and continues to live together with the country's white minority. They could have chosen justice based on retribution – an eye for an eye, on forgetting don't think about the past, just move forward into the future; or on restoration – granting amnesty in exchange for truth. This is how Tutu explains their choice:

> *We believe in restorative justice. In South Africa, we are trying to find our way toward healing and the restoration of harmony within our communities. If retributive justice is all you seek through the letter of the law, you are history. You will never know stability. You need something beyond reprisal. You need forgiveness.*[11]

The commission received over 7,000 applications for amnesty in exchange for truthful accounts of violations of human rights. Before finishing its work in 1998, nearly 2,000 people testified before the commission, and it received roughly 20,000 statements of rights abuses. In concluding the work of the TRC, Tutu said that, although he was "devastated by the depths of depravity that the process has revealed", he also had "been amazed, indeed exhilarated by the magnanimity and nobility of spirit of those who, instead of being embittered and vengeful, have been willing to forgive those who treated them so horribly badly."[12] The deputy chairman, Dr. Alex Boraine, said that perhaps the greatest contribution of the TRC toward social reconciliation was the recognition that 'reconciliation is not easy, is never cheap and is a constant challenge'. Although the process of bringing closure to the era of apartheid was messy and agonizing, it was effective in creating the foundation for a new beginning. Boraine explained that many people testified that appearing before the commission finally put an end to their 'nightmares of isolation' and that, for the first time since losing their loved ones, they could sleep at night. Yet others told of a 'broken heart which had been healed'.[13]

The strong sense of community found in South African culture helped to inspire this approach to reconciliation. In the African view, the community defines the person. The word for this is ubuntu which, translated roughly, means 'each individual's humanity is ideally expressed in relationship with others' or 'a person depends on other people to be a person'.[14] From this feeling for community emerged the non-violent means to move South Africa from racial separation and minority rule to integration and democracy.

A second step in the process of reconciliation is for the wrongdoer to offer a sincere public apology. Here are examples of important public apologies offered in recent years:[15]

- In 1988, an act of Congress apologized 'on behalf of the people of the United States' for the internment of Japanese Americans during World War II.

- In 1996, German officials apologized for the invasion of Czechoslovakia in 1938 and established a fund for the reparation of Czech victims of Nazi abuses.

- In 1998, the Japanese prime minister expressed 'deep remorse' for Japan's treatment of British prisoners during World War II.

A powerful example of a public apology and social healing is provided by the relationship between the Aboriginal people and the European settlers in Australia. On May 26, 1998, Australia commemorated its first 'Sorry Day' to express people's regret and shared grief about a tragic episode in Australian history – the organized removal of Aboriginal children from their families on the basis of race. Through much of this century, Aboriginal children were forcibly removed from their families with the aim of assimilating them into Western culture.[16] According to an indigenous council member, Patricia Thompson, Sorry Day provides a way for Australians to come to terms with their history and to come together to build a future on a foundation of mutual respect. Said Thompson, "What we want is recognition, understanding, respect and tolerance – of each other, by each other, for each other." In cities, towns, and rural centers, in schools and churches, people stopped their everyday activities to acknowledge this injustice. In addition, hundreds of thousands of Australians have signed the 'Sorry Books'.

The third step in reconciliation is restitution or the payment of

The strong sense of community found in South African culture helped to inspire this approach to reconciliation. In the African view, the community defines the person.

reparations. Archbishop Desmond Tutu gives a good explanation of the role of restitution when he says that, completing the process of reconciliation involves more than the recognition and remembering of injustice: "If you steal my pen and say 'I'm sorry' without returning the pen, your apology means nothing."[17] In cases like this, what is needed is restitution. Apologies create a truthful record. Restitution creates a new record. The purpose of reparation is to repair the material conditions of a group so as to restore some balance or equality of power and material opportunity.[18]

Beyond reconciliation is the day-to-day reality of former antagonists living together. One of the most notable examples of successful reconciliation within the recent past is the shift in the relationship between the United States and Germany and Japan. World War II began in the age of total warfare, when massive civilian causalities were the norm; it could have taken many generations to heal the psychological wounds from that war. Yet, within a few decades, the United States and its bitter enemies from the war, Germany and Japan, became peaceful allies – clear examples of successful reconciliation culminating in renewed relationships and mutual respect. Other important examples of reconciliation include the ongoing peace process in the Middle East, certainly one of the most volatile regions in the world, and in Northern Ireland, where the process of reconciliation seems poised to overcome centuries of separation and conflict.

As these examples make clear, with authentic reconciliation – with listening, apologizing, and restoring – the suffering of the past does not need to stand in the way of future progress.

... with authentic reconciliation – with listening, apologizing, and restoring – the suffering of the past does not need to stand in the way of future progress.

The Cost of Kindness

The cost of compassion is far less than we might think. The world does have the material resources for all of us to live together sustainably. We could begin by eliminating the worst aspects of poverty – a fundamental requirement, I believe, for an evolutionary bounce to occur. As the United Nations 1998 Human Development Report concludes, we have 'more than enough' resources to accomplish this.[19] To make this point, the report presents these stark contrasts:

- To achieve universal access to water and sanitation, the estimated additional annual cost is $12 billion, which is what is spent on perfumes in Europe and the United States each year.

- To achieve universal basic health and nutrition, the estimated additional annual cost is $13 billion, which is $4 billion less than annual expenditures on pet foods in Europe and the United States.

- The world's spending priorities are further reflected in these figures: Annual expenditures on business entertainment in Japan amount to $35 billion; on cigarettes in Europe, $50 billion; on alcoholic drinks in Europe, $105 billion; and on military spending in the world, $780 billion.

The Human Development Report concludes, 'advancing human development is not an exorbitant undertaking'. The added bill to provide universal access to basic services – education, health, nutrition, reproductive health, family planning, safe water, and sanitation – is estimated to be an additional $40 billion per year.[20] This is less than one-tenth of one-percent of world income. As the report notes, this is 'barely more than a rounding error'.

Given that we can easily afford to eliminate the worst forms of poverty, what are we doing about it? The report states that development aid is now at its lowest level since the U.N. started keeping statistics. Donor countries allocate an average of only 0.25 percent (one quarter of one percent) of their total GNP to development assistance for poorer nations. The United States is the stingiest developed nation in terms of the proportion of total wealth that it donates.[21]

The resources exist to make a dramatic improvement in the quality of life for a majority of humanity and to begin a process of reconciliation between the rich and the poor. Instead of trickle-down development from the wealthy to the poor, we should launch a bottom-up approach that directly targets the poor and the voiceless.[22] If we use equity, simplicity, and cooperation as our guideposts, we have the resources to sustain all of humanity into the foreseeable future. As Gandhi said, "We have enough for everyone's need, but not for everyone's greed."

We cannot achieve our maturity if we remain divided into a minority that has great wealth and a majority that is consigned to absolute poverty. We need to create a future of mutually assured development where progress leaves no one behind and also strengthens the ecosystems on which our common future depends. We could create something akin to the Marshall Plan, which restored Europe after World War II. The entire world could be united in establishing a foundation of sustainability. Given intelligent designs for living lightly and simply, a decent standard and manner of living could vary depending on local customs, ecology, resources, and climate. Within this diversity, if the human family saw its collective development as its central enterprise, the world would have a strong foundation for an evolutionary bounce.

Archbishop Desmond Tutu said that you can immediately tell when you enter a happy home: "You don't have to be told; you don't have to see the happy people who live there. You can feel it in the fabric, the air."[23] In a similar way, he says, we have it in our power to create a cultural atmosphere on Earth that is infused with kindness, joy, laughter, truth, and love. If we can bear witness to the reservoir of unresolved pain that has accumulated through history, we will release an enormous store of pent-up creativity and energy. Instead of mobilizing around enemies, we could release our collective energy in building a sustainable and promising future.

We cannot achieve our maturity if we remain divided into a minority that has great wealth and a majority that is consigned to absolute poverty.

1. Pierre Teilhard de Chardin, *The Future of Man*, New York: Harper & Row, 1964, p.57.
2. Pitirim Sorokin, *The Ways and Power of Love*, Chicago: Henry Regnery Co., 1967 (originally copyright 1954).
3. Jack Kornfield, 'The Path of Compassion: Spiritual Practice and Social Action', in *The Path of Compassion*, Fred Eppsteiner, ed., Berkeley, CA: Parallax Press, 1988, p.29.
4. Pitirim Sorokin, *The Ways and Power of Love*, Chicago: Henry Regnery Co., 1967 (originally copyright 1954), p.71.
5. This description is drawn primarily from: Sorokin, Op. Cit., p.67 and Eknath Easwaran, *The Compassionate Universe*, Petaluma, CA: Nilgiri Press, 1989.
6. Gitanjali Kolanad, *Culture Shock!* India, Portland, OR: Graphic Arts Center Publishing, 1994, p.23.
7. Sorokin, Op. Cit., p.68.
8. Ibid., p.110.
9. Ibid., p.69.
10. Donald Shriver, Jr., *Forgiveness in Politics*, New York: Oxford University Press, 1995., p.7
11. Desmond Tutu quoted in Terry Tempest Williams, 'Two Words', Orion, Great Barrington, MA, Winter 1999, p.52.
12. Archbishop Desmond Tutu, 'A Message from the Chairperson', *Truth Talk: The Official Newsletter of the Truth and Reconciliation Commission*, South Africa, July 1998.
13. Dr. Alex Boraine, 'A message from the Deputy Chairperson of the TRC', *Truth Talk: The Official Newsletter of the Truth and Reconciliation Commission*, South Africa, July 1998.
14. Michael Battle, *Reconciliation: The Ubuntu Theology of Desmond Tutu*, Ohio: The Pilgrim Press, 1997, p.39.
15. These examples were drawn, in part, from: Emily Mitchell, 'The Decade of Atonement', *Index on Censorship*, May/June 1998, London (and reprinted in the Utne Reader, March-April, 1999, pp.58-59).
16. John Bond, 'Aussie Apology', *Yes! A Journal of Positive Futures*, Bainbridge Island: WA, Fall 1998, p.22.
17. Ibid., p.224.
18. Eric Yamamoto, *Interracial Justice: Conflict and Reconciliation in Post-Civil Rights America*, New York University Press, 1999.
19. *Human Development Report 1998*, United Nations Development Programme, New York. Oxford University Press, 1998, p.37
20. Ibid.
21. Slobodan Lekic, 'Rich Nations Grow More Stingy With Poor Nations', *San Francisco Chronicle*, October 17, 1997, World Section, p.A14.
22. Glenys Kinnock, 'One World', in E. and D. Shapiro, eds., *Voices from the Heart*, New York: Tarcher/Putnam, 1998, p.122.
23. Desmond Tutu, 'Becoming More Fully Human', in Eddie and Debbie Shapiro (eds.), *Voices from the Heart*, New York: Jeremy Tarcher/Putnam, 1998, 277.

Duane Elgin, MBA and MA is an internationally recognized speaker and author. His books include: *The Living Universe: Where Are We? Who Are We? Where Are We Going?* (2009), *Promise Ahead: A Vision of Hope and Action for Humanity's Future* (2000), *Voluntary Simplicity: Toward a Way of Life that is Outwardly Simple, Inwardly Rich* (2010, 1993, 1981), and *Awakening Earth: Exploring the Evolution of Human Culture and Consciousness* (1993). In 2006, Duane received the international 'Goi Peace Award' in recognition of his contribution to a global "vision, consciousness, and lifestyle" that fosters a "more sustainable and spiritual culture." Duane's website is: **www.duaneelgin.com**

Recent discoveries in cardiology reveal the heart to be neurologically far more complex than previously believed – shedding new scientific insights on emotional energies and spiritual practices of heart prayer.

The Intelligent Heart

Michael Stubberup and Matias Ignatius

Throughout history humanity has known about the importance of the heart when it comes to emotional development, transformation of consciousness and existential wisdom. The great spiritual traditions in East and West practice techniques based on the connection between respiration and the heart rhythm. Centuries of experience with inner states have given man profound knowledge about the coherence among body, mind and spirit – and the role of the heart in relation to this.

The knowledge gathered through centuries of experience within the great wisdom traditions builds on experiments, observation, repetition and the ability to reproduce specific inner states of the mind. One could therefore argue that a science of inner psychological states has been developed – a science based on centuries of empirical evidence. The Western world has only acknowledged that which can be objectively measured as science. But with the development of new technologies such as certain types of brain scanners, and the rise of neuroscience and the establishment of neurophenomenology, we are now approaching a situation in which inner and outer science may become two sides of the same question.

The separation of inner and outer is visible for instance in language. Heart is called 'kardia' in Greek. We know the use of kardia from modern medicine, for example, cardiogram (a curve that shows the heart activity) or cardiologist (heart specialist). So, it is a scientific term, which in different ways refers to the physical heart. However, the original Greek-Byzantine word has a far more comprehensive meaning. The heart is in this context not only understood as a physical organ but also as the centre of man's spiritual life; the heart is understood as the deepest and most authentic Self, through which the mystical union of the divine and the human can take place. In recent years one place in particular has worked towards a meeting of inner and outer science into focus: The HeartMath Institute in Boulder

Creek, California.

In the following, we will give a brief introduction to the story behind the HeartMath Institute, aspects of the latest cardiology science and the incipient meeting between modern science and the practice of heart prayer.

The HeartMath Institute

The basis for The HeartMath Institute was developed over a period of more than ten years. During the late seventies and the eighties the founder of HeartMath Institute, Doc Childre, studied the knowledge about the science of heart that was available at the time. Over the years he managed to build a network of scientists. In 1991 he founded HeartMath as a non-profit organization with the purpose of uniting the science of the heart with the field of education. The intention was to develop a set of techniques that would be accessible to anyone and would have the potential of creating improved physical and emotional balance and health. Today HeartMath is involved in a wide range of activities from scientific initiatives to training and educational programs for students as well as professional training programs for companies and organizations.

Neurocardiology: A brain Within the Heart

During their research in the sixties and seventies on the communication between the brain and the heart, John and Beatrice Lacey found that the heart communicates information to the brain in ways that are crucial for how we understand and respond to the world. Following extensive research, another pioneer of neurocardiology, Dr. J. Andrew Armour, in 1991 developed the concept of an actual neural network within the heart. He described his scientific discovery as follows:

> ...that the heart has a complex intrinsic nervous system that is sufficiently sophisticated to qualify as a 'little brain' in its own right.

The brain of the heart functions as an independent network consisting of different types of neurons, neurotransmitters, proteins, etc. This network enables the heart to act independently from the cranium brain, to sense and feel, learn and remember.

Connections Between the Heart and the Brain

The heart and the brain communicate in a number of different ways, which researchers at HeartMath Institute describe it this way:

> ...we have now learned that communication between the heart and the brain is actually a dynamic, ongoing, two-way dialogue, with each organ continuously influencing the others function.

This communication is facilitated in four ways: a) energetically via the electromagnetic field, b) neurologically via the nervous system, c) biochemically via hormones and neuro transmitters and d) biophysically via the pulse waves in the arteries.

The electromagnetic signal from the heart arrives in the brain immediately, while a series of neural signals start arriving after eight milliseconds. After approximately 240 milliseconds the blood pressure wave arrives, which subsequently synchronizes the neural activity – particularly in relation to the brain's alpha waves.

The brain communicates with the heart through the autonomic nervous system (ANS), which consists of two sub-systems: the sympathetic nervous system, which prepares the body for activity by speeding up the heart rhythm, and the parasympathetic nervous system, which prepares the body for rest by lowering the heart rhythm. The autonomic nervous system regulates approximately 90% of our bodily functions. The activity in the two parts of ANS creates changes in the heart rhythm and this is measured as heart rate variability (HRV).

Emotions and Heart Rhythm

As early as in ancient Greece it was commonly assumed that emotions and thinking were separate processes. In the age of enlightenment Descartes subscribed to this notion as illustrated by his familiar statement: "I think, therefore I am." However, as the neurologist Antonio Damasio proposed in his book Descartes' Mistake (1994), emotion and thinking are commonly closely connected and emotions in general are a prerequisite for decision making.

Researchers at HeartMath Institute carried out different experiments measuring emotional states and their direct connection with the heart. They did this by analyzing oscillations with ECG, which revealed reactions, albeit small ones, to highly negative emotions. Heartbeat per minute was also investigated but no significant correlation with emotional states was found. However, when the researchers began investigating changes between individual heartbeats, an interesting breakthrough was made. They discovered that the changes from one heartbeat to the next reflected changes in the emotional state. This discovery was first published in the *American Journal of Cardiology* in 1995 by McCraty and colleagues: "We have found that heart rate variability patterns are extremely responsive to emotions, and heart rhythms tend to become more ordered or coherent during positive emotional states." In subsequent studies it was found that positive and constructive emotions like kindness, gratitude, solicitude and love create soft and structured patterns in the heart rhythm (coherence). On the other hand, negative emotions like anger, frustration and anxiety create disorganized and chaotic heart rhythm patterns. Most people experience that emotions come about and have an impact without the person being able to do much actively to change the emotional condition or its development.

... positive and constructive emotions like kindness, gratitude, solicitude and love create soft and structured patterns in the heart rhythm (coherence).

With inspiration from different spiritual traditions HeartMath Institute developed a range of different self-development techniques. An example is the Quick Coherence technique, which was designed to teach people how to actively train and develop coherent heart rate patterns. Through this type of exercise it is possible quite easily to access a relaxed inner condition of emotional balance and greater mental clarity.

Quick coherence:

1 First, focus your attention on the area around your heart.

2 Next, breathe as if your breath is flowing in and out through the center of your chest.

3 As you breathe try to find and remember an appreciative situation.

Heart Coherence and Computer Training

In 2000 the HeartMath Institute presented the Freeze Framer, a computer programme designed to help people train the ability to attain a more coherent condition. Since then the program has won a number of awards for its usefulness and precision. Freeze Framer is a bio-feedback system that connects the person via a sensor on either the finger or earlobe to the computer. The sensor communicates the pulse and the electromagnetic signal to the computer, which continuously illustrates the current degree of coherence between the heart and the autonomic nervous system. The level of coherence is shown on the computer screen in real time, so that one is able to follow the development of inner balance directly as it changes while doing the exercise. Imbalance is shown as sharply jagged curves without much structure, while a state of more coherence and balance is depicted as gently sloping and structured curves resembling sinus waves. Furthermore, the program shows three colored columns – a red, blue and green – where red signifies imbalance, green means a high degree of coherence and blue indicates a lesser degree of balance. With these two illustration tools – the continuous curve and the three columns – it is possible to read one's degree of balance or imbalance during the exercise and enabling one to respond to the feedback on the screen when distractions occur.

It is indeed interesting to find that exercises which are very similar to basic exercises in many of the great spiritual traditions, can now be found in some of the most advanced scientific environments in the USA.

In our experience Freeze Framer can be used as an inspiring supplement to a self-development or meditation practice, and as a modern tool for working towards the maturation of the heart. It is a fascinating experience to see one's inner state depicted in real time while working with breathing and heart exercises. The Freeze Framer is surprisingly sensitive and precise in registering distraction and lack of focus. This makes it an interesting tool because it can distinguish between focused presence and tenser, distracted states. If one becomes too eager or ambitious, this will immediately show on the screen as disorganized patterns. On the other hand, if you are too imprecise or insufficiently focused, the feedback will show a lack of coherence as well. In this way, the ability to build a sharp but at the same time gentle and not overly ambitious focus is effectively reflected through the program. By using Freeze Framer on a regular basis, it is possible to learn how to access a more coherent state and thereby how to actualize a wider range of the heart's profound, natural qualities. In time you get to know some of your inner habitual patterns. And you realize how little it actually takes for an almost optimal coherent state to be disturbed by an emotion or thought, and at the same time how relatively difficult it is to regain such a state of coherence.

Psychophysiological Coherence

The heart is the most powerful generator of rhythmic information patterns in the body. With each heartbeat a complex pattern of neurological, hormonal and electromagnetic information is sent to the brain and the rest of the body.

When the pattern of the heart rhythm is coherent, the neurological information that is sent to the brain is strengthened. This coherent pattern, which is called psychophysiological coherence by the HeartMath Institute, is a total picture of a series of interrelated harmonious bodily and psychological conditions. This coherent condition is largely dependent on a balance and synchronization of cognitive, emotional and physiological processes. This is commonly subjectively experienced as increased mental clarity and spontaneous creativity. One of the ways in which psychophysiological coherence is transmitted from the heart to the brain, is through the heart centre (medula) in the brainstem on to the thalamus and amygdala. These areas are closely connected to the frontal lobes, where reason and emotions are integrated and form the basis of choices and decisions.

The role of the amygdala is especially interesting in relation to working with therapy and trauma. The amygdala is the centre where behavior and immunological and neuroendocrine responses to threats from the environment are coordinated. Moreover, it is the emotional memory center of the brain. If there is the slightest correspondence between the traumatic memory and the incoming stimulus, the amygdala immediately sends a signal to the autonomic nervous system and the body's fight/flight mechanism is activated. What is particularly interesting is the direct connection between the amygdala and the heart rhythm: The activity in the core cells of amygdala syncronises with the heart rhythm. All indications are that regular heart coherence over time is the physiological correlate of the harmonization process of trust and emotions, which takes place between therapist and client in deep therapeutic relations.

The heart is the most powerful generator of rhythmic information patterns in the body.

Bioelectric Communication

The heart wave sends out a rhythmic pattern, which reflects the emotions one feels. Emotional states are sent out to other people, among other things through the electromagnetic field generated by the heart. The electric field of the heart is 60 times more powerful than the brain's and the magnetic field of the heart is 5000 times more powerful than the brain's. The brain's energy field reaches no more than a few centimetres away from the body, while the energy field of the heart has been measured as reaching more than three meters away from the body. This means – which in a sense isn't surprising at all – that we are strongly affected by the electromagnetic energy that radiates from another person's heart. This phenomenon has been thoroughly investigated at The HeartMath Institute. In one experiment a subject sits on a chair and performs a heart coherence exercise while another person sits a few metres from him in a relaxed state while the EEG is being recorded. The fascinating finding from this experiment is that the coherent heart waves of the first person can be seen directly in the other person's brain waves. In another experiment a boy and his dog were tested. The heart rhythm pattern of the boy and the dog was measured when they were together and separated. Before and after they were together, the heart rhythm patterns were disorganized and chaotic. When they were together on the other hand, the patterns were not only coherent but even synchronized.

The brain's energy field reaches no more than a few centimetres away from the body, while the energy field of the heart has been measured as reaching more than three meters away from the body.

Prayer and HeartMath

Already from around 300AD in the Hesychastic tradition of the Eastern Orthodox Church the heart was considered the center of consciousness. The spiritual system and a range of inner tools focusing on the heart were based upon this fundamental notion. Gradually, the prayer of the heart came to be the central practice. It is a simplified prayer in which God is contacted through repetition of a single phrase while keeping the focus in the heart. Over the course of centuries, Hesychastic practices were developed centered on different bodily sites and functions, respiration and heart rhythm being the most important. In one of the most widely known texts from the Eastern Church, a pilgrim describes his practice this way: "Later I began to practise the Jesus prayer in and out through the heart coordinated with the breathing as Gregor from Sinai and also Kallistos and Ignatios teach us. That means when I was inhaling I was looking with the minds eye into the heart while I was thinking and saying 'Lord Jesus Christ' and when I was exhaling, 'have mercy on me.'"

Within the tradition this exercise is called the Supreme Gate of the Heart. Comparison of this quotation with the Quick Coherence from HeartMath Institute, reveals a striking resemblance in both structure and contents. In both systems there is a) focus on the heart, established by b) breathing in and out through the heart and c) coordination of respiration by means of visualization of a 'high' or positive and meaningful quality. The crucial difference lies in the different context of the two systems. The prayer of the heart is developed within a religious context in which there is a metastructure in the form of contacting the transcendent ('heaven', Christ). This is not the case in the context of HeartMath Institute.

As science is now beginning to discover and verify what has been known within the spiritual traditions for thousands of years, there may be an opportunity for spreading knowledge about such valuable inner practices in an undogmatic way. These may prove to be extremely fruitful for people's professional and personal development, as well as in relation to basic existential conditions.

Michael Stubberup and Mathias Ignatius, both from Denmark, work with a group of psychologists and teachers exploring consciousness practices with roots in both traditional heart prayer and the neurocardiology of the Hearth Math Institute. They also work with 'children's wisdom', helping children to find their inner core, focus and relax by introducing simple techniques in school.

Karambu Ringera tells the moving story of how peace circle dialogues were set up in Kenya and the healing they have brought. She describes why women have a vital role to play in peace work and politics.

Peace Circle Dialogues: I Am Because You Are

Karambu Ringera

When I arrived at the Internally Displaced Peoples (IDP) camp at Nakuru, Kenya in January 2008, nothing prepared me for the sight of the effects of war in my beloved Kenya. As I walked about surveying the violence, I watched as a woman sold her tomatoes, unbothered by the presence of a dead body near her stand. There was so much death throughout the streets. I am shocked at what we have become as a nation – we are so removed from our humanity, we fail to see that the 'other' being butchered here are fellow Kenyans.

In the camps, my heart bled for the women and children I met. For food, they have a mug of porridge in the morning, no lunch, and a dinner so meagre 'it is meant to keep the soul alive', as one old woman told me. Mothers have to forgo their own meals to feed their children. Girls in the camp are known to exchange sex for food.

Rapes occur regularly; women are told to watch over their girls, to not go to the toilets at night. One woman told me: "At night the men scream, and as the women run away in panic, they chase the women and girls and rape them."

HIV and AIDS is spreading like wildfire throughout the camps. The Kenya Red Cross Society (KRCS) medical team hands out condoms each day, and every morning used condoms are found all over the camp. HIV positive mothers have no food to feed their children; instead, they must breastfeed their babies to keep them from screaming out in hunger.

At Nakuru, there is no education for the children. Women begged for assistance in locating educational resources and materials for their children. They wanted to know where to take their children for schooling; how they could leave the camp and reach their relatives; how they could earn a living. Because I am from Meru, nine hours by car from Nakuru, and did not have the capacity to help, I directed the women to the Kenyan Red Cross services. At one point, a very determined woman told me: "You are not leaving me here!" Today I live with her and three of her children in my house. At the time of writing, she has no idea where her first born son is. She has not seen him since the day each one of them ran in different directions to escape the violence.

What went wrong after the elections? Many reasons have been suggested for the outbreak of post election violence in Kenya after the 2007 general elections that claimed over 2,000 lives and property loss of millions of Kenyan shillings. Although the announcement of the presidential results triggered the violence, there are several far-reaching root causes of the violence, including the colonial legacy that left a deeply divided people, disputed land ownership, and weak democratic institutions.

Today, Kenya – a nation that leads peace missions to other countries – is looking to outside peace mediators to quell the rising tide of anarchy. To date, few intervention strategists understand that women and children experience conflicts differently than men. In times of crisis, we must recognize that the assistance that meets the needs of men does not always meet the needs of women and children.

Peace Circle Dialogues

As I surveyed the IDP camp, I quickly realized it was important to find some way to empower these women, and to encourage the development of peaceful leaders. As a peace and social change activist, I felt there was a need for focused peace dialogues to help address the diverse needs of women and children. While many organizations are providing counseling to the displaced, women are left wondering how they will live alongside those who destroyed their homes and businesses. I felt that organizing peace dialogues was one way to start addressing these issues, a place to start mending this pain.

Organizing informal peace circle dialogues, I began by encouraged the women to share their stories. One Friday, the women and I sat in a circle. Some told me of painful family issue – husbands forcing wives to leave their home because they were from a different ethnic group; mothers forced to take their children elsewhere because they have the blood of an unwanted ethnic group. One woman told me of an elderly woman who was gang raped and then ripped open because the gangsters wanted to 'see where they had been'. The woman died.

In these gatherings, I offered three statements to help guide our dialogue. I asked women to complete the statements: 1. "I have peace when..."; 2. "Peace for me is..."; and, 3. "In the name of peace I commit to...". The answers to these questions were varied. Some women said they have peace when they pray and read the Bible; others when they are able to provide food, shelter and health to their families; still others when they can provide education to their children.

The women also gave varying definitions of peace: Peace is when the conflict and violence ends; peace is when all the children in the camp can return to school; peace is when there is help for those with HIV/AIDS, widows, single mothers and orphans. The women continuously told me that education is the only hope for their children because all of their property is gone and they can no longer support their families. Finally, I asked the women to commit to praying for peace; to supporting those in need; to encouraging each other to hope in God.

A Way Forward

The violence in my beloved Kenya and the time I spent in the IDP camps has reaffirmed my desire to participate in healing the broken cord that joins me to my sisters and brothers, no matter their ethnic origin. This healing begins with women. Just as women have been absent in Kenyan politics since 1963 when we attained independence, in the current initiative to end the political crisis in Kenya, women have been sidelined. We must make concerted efforts to include Kenyan women in peace-building initiatives if we are to see progress.

Convinced that peace dialogues are a first step in the way forward, I began creating a peace-training program under the Institute for Nonviolence and Peace (INPEACE), a program that was launched in 2005 at the Women's Congress held in Nairobi that same year. As a short-term response to the crisis, the program is conducting peace trainings for women and youth in IDP camps. Beyond this, the training of leaders, especially politicians, in peace discussions is important so that they can share this knowledge with their constituents. I recently successfully appealed to local organizations in Kenya for finances to support this program both in the camps and among parliament leaders.

There are, of course, many things that must happen to quell the violence. Currently, the former UN Secretary General-led peace initiative has seven men and only two women, both of whom are politicians. Human rights organizations, civil societies, FBOs, CBOs and NGOs have no representation in this initiative at all. Although this is a political crisis said to need a political solution, other civil society entities need to be included in the mediation effort.

Most importantly, women must be represented because they have unique experiences of violence that need to be highlighted and recognized at the mediation table. In lieu of the current events in Kenya, there is an urgent need to implement Resolution 1325, the 2000 UN Security Council resolution that urges a gender perspective in conflict prevention and conflict resolution. It also calls upon states and actors to ensure women's full participation in peace processes.

Africa is my Sanctuary

Africa, a sanctuary? A safe haven, shelter, protection, place of safety, harbor, retreat, and asylum. 'Home' comes to mind when I think of sanctuary. And

Africa epitomizes the concept of 'home'. Home, a place that is supposed to be 'safe' yet is not. It really is a constant reminder that there is no safe space anywhere. Africa is viewed by outsiders as a place of chaos, yet all spaces have their own types of chaos. Focusing on Africa or any other place at war blinds us from looking at our own spaces of chaos; it makes us always look outward, (un)safely located in our homes, instead of looking inward, where our own woundedness dwells.

I have been one who says that Africa is my sanctuary. Women in war-torn Juba in Southern Sudan said to me "I do not know what caused the war." Another said, "We woke up one morning, and there was war." When I asked them what they did for peace, they said: "I prayed. Every day I knelt down and lifted my hands up to God and prayed for peace." And I remember thinking: That is all they did? Just prayed? The women of Kenya also kept saying they had left everything in the hands of God. One even said to me that she had found forgiveness in her heart for the killers and destroyers of their homes. She said: "Once I really forgave, I was free in my heart. Nowadays I sleep like a baby."

Upon reflection I realized that these women not only gave everything they had, but they did something more: they prayed. This was something they could offer for peace in their country – they surrendered to a greater power that will never abandon them no matter what. To pray, to surrender; at first I thought that this was a measure of the women's helplessness, and then I rethought. These women were not helpless, they never gave up – they could pray. And their praying is one thing no one could take away from them.

These women were not helpless, they never gave up – they could pray. And their praying is one thing no one could take away from them.

Today's world needs leaders who inspire and empower others to step into their own power. Change can only come from within – we are all leaders, and if we endeavor to empower others, we can create strong partners for change in our communities.

I have a belief that I must learn how to create peace in my own heart first, because unless I am at peace in my self, and unless I know myself, then it will be difficult to know and understand another and offer peace or even create peace for self and other. In the end, I recognize I am the other and the other is me. In a continent torn by retaliatory strikes and conflicts premised on ethnic and religious differences, we must continue to make efforts toward this goal – recognizing that the other is me and I am the other:

"I am because you are."

Karambu Ringera, PhD, is the founder and president of International Peace Initiatives (IPI) – a US and Kenya-based organization that promotes peace and supports efforts to mitigate the effects of war, poverty and discrimination at the grassroots level. She received her PhD in Human Communications from the University of Denver, Colorado, and holds degrees from the University of Nairobi, Natal University in South Africa, and Iliff School of Theology in Denver. In 2007, Karambu ran for Kenyan Parliament in North Imenti, placing in sixth out of fifteen all-male candidates.

Wangari Matthai describes how restoring cultural values that have been destroyed by colonialism can affect a deep healing, not only of people but also of the environment.

The Cracked Mirror

Wangari Maathai

Mount Kenya is a World Heritage Site. The equator passes right on its top, and it has a unique habitat and heritage. Because it is a glacier-topped mountain, it is the source of many of Kenya's rivers. Now, partly because of climate change and partly because of logging and encroachment through cultivation of crops, the glaciers are melting. Many of the rivers flowing from Mount Kenya have either dried up or become very low. Its biological diversity is threatened as the forests fall.

"What shall we do to conserve this forest?" I asked myself. As I tried to encourage women and the African people in general to understand the need to conserve the environment, I discovered how crucial it is to return constantly to our cultural heritage. Mount Kenya used to be a holy mountain for my people, the Kikuyus. They believed that their God dwelled on the mountain and that everything good – the rains, clean drinking water – flowed from it. As long as they saw the clouds (the mountain is a very shy mountain, usually hiding behind clouds), they knew they would get rain.

And then the missionaries came. With all due respect to the missionaries (they are the ones who really taught me), in their wisdom, or lack of it, they said, "God does not dwell on Mount Kenya. God dwells in heaven."

We have been looking for heaven, but we have not found it. Men and women have gone to the moon and back and have not seen heaven. Heaven is not above us: it is right here, right now. So the Kikuyu people were not wrong when they said that God dwelled on the mountain, because if God is omnipresent, as theology tells us, then God is on Mount Kenya too. If believing that God is on Mount Kenya is what helps people conserve their mountain, I say that's okay. If people still believed this, they would not have allowed illegal logging or clear-cutting of the forests.

After working with different Kenyan communities for more than two decades, the Green Belt Movement (GBM), which I led until joining the new Kenyan government in January 2003, also concluded that culture should be incorporated into any development paradigm that has at its heart the welfare of the people. The Green Belt Movement's mission is mobilising community consciousness for self-determination, equity, improved livelihood security and environmental conservation – using trees as the entry point. When we began, we believed that all that was needed was to teach people how to plant trees and make connections between their own problems and their degraded environment.

But in the course of struggles to realise GBM's mission and vision, we realised that some of the communities had lost aspects of their culture which had actually facilitated the conservation of that beautiful environment which the first European explorers and missionaries recorded in their diaries and textbooks.

Culture and our Environment

Culture is an important part of humanity. Development agencies, religious leaders, and academic institutions are increasingly recognising its central role in the political, economic and social life of communities. A focus on culture is important to environmentalists as well as to traditional communities. Too often, when we talk about conservation, we don't think about culture. But we human beings have evolved in the environment in which we find ourselves. For every one of us, wherever we were, the environment shaped us: it shaped our values; it shaped our bodies; it shaped our religion. It really defined who we are and how we see ourselves.

Cultural revival might be the only thing that stands between the conservation or destruction of the environment, the only way to perpetuate the knowledge and wisdom inherited from the past, necessary for the survival of future generations. A new attitude toward nature provides space for a new attitude toward culture and the role it plays in sustainable development: an attitude based on a new understanding – that self-identity, self-respect, morality, and spirituality play a major role in the life of a community and its capacity to take steps that benefit it and ensure its survival.

Until the arrival of the Europeans, communities had looked to Nature for inspiration, food, beauty and spirituality. They pursued a lifestyle that was sustainable and that gave them a good quality of life. It was a life without salt, soap, cooking fat, spices, soft drinks, daily meat, and other acquisitions that have accompanied a rise in the 'diseases of the affluent'. Communities that have not yet undergone industrialisation have a close connection with the physical environment, which they often treat with reverence. Because they have not yet commercialised their lifestyle and their relation with natural resources, their habitats are rich with local biological diversity, both plant and animal.

However, these are the very habitats that are most at threat from globalisation, commercialisation, privatisation, and the piracy of biological materials found in them. This global threat is causing communities to lose their rights to the resources they have preserved throughout the ages as part of their cultural heritage. These communities are persuaded to consider their relationship with Nature primitive, worthless, and an obstacle to development and progress in an age of advanced technology and information flow.

The Wounds of Colonialism

During the long, dark decades of imperialism and colonialism from the mid-nineteenth century to the mid-twentieth century, the British, Belgian, Italian, French and German governments told African societies that they were backward. They told us that our religious systems were sinful, our agricultural practices inefficient, our tribal systems of governing irrelevant, and our cultural norms barbaric, irreligious, and savage. This also happened with the Aborigines in Australia, the Native Americans in North America, and the native peoples of Amazonia.

Of course, some of what happened, and continues to happen, in Africa was bad and remains so. Africans were involved in the slave trade; women are still genitally mutilated; Africans are still killing Africans because they belong to different religions or ethnic groups. Nonetheless, I for one am not content to thank God for the arrival of 'civilisation' from Europe, because I know from what my grandparents told me that much of what went on in Africa before colonialism was good.

There was some degree of accountability to people from their leaders. People were able to feed themselves. They carried their history – their cultural practices, their stories and their sense of the world around them – in their oral traditions, and that tradition was rich and meaningful. Above all, they lived with other creatures and the natural environment in harmony, and they protected that world.

Agriculture, democracy, heritage, and ecology are all dimensions and functions of culture. Agriculture is the way we deal with seeds, crops, harvesting, and processing and eating. One result of colonialism was the loss of indigenous food crops such as millet, sorghum, arrowroot, yam, and green vegetables, as well as livestock and wildlife. Like culture itself, the possession of cattle as a sign of wealth or the growing of one's own food were trivialised by colonisers as indicators of a primitive mode of living. Loss of indigenous food and the methods to grow it have contributed to food insecurity at the household level and diminishment of local biological diversity.

People without culture feel insecure and are obsessed with the acquisition of material things, which give them a temporary security that itself is a delusional bulwark against future insecurity. Without culture, a community loses self-awareness and guidance, and grows weak and vulnerable. It disintegrates from within as it suffers a lack of identity, dignity, self-respect and a sense of destiny.

Finding our Own Mirror

By the end of the civic and environmental seminars organised by the Green Belt Movement, participants feel the time has come for them to hold up

Without culture, a community loses self-awareness and guidance, and grows weak and vulnerable. It disintegrates from within as it suffers a lack of identity, dignity, self-respect and a sense of destiny.

their own mirror and find out who they are. This is why we call the seminars kwimenya (self-knowledge). Until then, participants have looked through someone else's mirror – the mirror of the missionaries or their teachers or the colonial authorities who have told them who they are and who write and speak about them – at their own cracked reflections. They have seen only a distorted image, if they have seen themselves at all!

There is enormous relief and great anger and sadness when people realise that without a culture not only is one a slave, but one has actually collaborated with the slave trader, and that the consequences are long-lasting. Communities without their own culture, who are already disinherited, cannot protect their environment from immediate destruction or preserve it for future generations. Since they are disinherited, they have nothing to pass on.

A new appreciation of culture can give traditional communities a chance, quite literally, to rediscover themselves, and to revalue and reclaim their culture. This is no trivial matter of reviving pottery or dancing, or whatever limited ideas of indigenous culture some Westerners may still have.

Of course, no one culture is applicable to all human beings who wish to retain their self-respect and dignity; none can satisfy all communities. Humanity needs to find beauty in its diversity of cultures and accept that there will be many languages, religions, attires, dances, songs, symbols, festivals and traditions. This diversity should be seen as a universal heritage of humankind.

Cultural liberation will only come when the minds of the people are set free and they can protect themselves from colonialism of the mind. Only that type of freedom will allow them to reclaim their identity, self-respect and destiny. Only when communities recapture the positive aspects of their culture will people relearn how to love themselves and what is theirs. Only then will they really appreciate their country and the need to protect its natural beauty and wealth. And only then will they have an understanding of the future and of generations to come.

© Friends of the Green Belt Movement, North America, www. **greenbeltmovement.org**

This article was first published in *Resurgence Magazine*, November 11, 2004.

Ecologist Wangari Maathai won the 2004 Nobel Peace Prize for her years of work with women to reverse African deforestation. Wangari Maathai went to college in the United States, earning degrees from Mount St. Scholastica College in Atchison, Kansas (1964) and the University of Pittsburgh (1966). She returned to Kenya and earned her PhD from the University of Nairobi (1971), then worked as a professor in their department of veterinary medicine. In 1976, Maathai began promoting a tree-planting program to reverse deforestation and provide firewood for Kenyan women. Soon known as the Green Belt Movement, the program led to the planting of millions of trees and made Wangari Maathai a major political figure in Kenya. In 1997 she ran unsuccessfully for president and for a seat in Parliament, but in December of 2002 she was elected to Parliament, and in 2003 she was appointed by President Mwai Kibabi to the Ministry of Environment, Natural Resources and Wildlife. She won the Nobel Peace Prize in 2004, with the Nobel committee citing 'her contribution to sustainable development, democracy and peace'. She was the first African woman to win a Nobel. She died in 2011.

Sister Lucy Kurien describes her remarkable Maher project, an inspiration for oppressed women everywhere. Maher has provided refuge and rehabilitation to more than 2,000 abused women in India.

Maher – Rising to New Life: Interview with Lucy Kurien

William Keepin

The Maher project located in and around Pune, India was originally founded as a refuge for battered women in 1997. Maher has grown into a thriving community and network of satellite communities engaged in highly innovative social, ecological, and spiritual work in the Pune region. Maher is an interfaith and intercaste community, where people of all castes and all religions live and thrive together in a joyous and loving community as they care for downtrodden and destitute women and children. Maher serves as a powerful beacon of hope in a society that has been devastated by an apartheid of rigid caste distinction, religious division, and gender oppression – a remarkable achievement in the highly patriarchal Indian society. The interview here is with Sister Lucy Kurien, the founding director of Maher.

What led you to found the Maher project?

It was actually in the year 1991, when I was working at a place called HOPE in Pune. One day, a pregnant woman from an adjacent apartment building came to my convent, asking for shelter. She told me that her husband, a chronic alcoholic, was threatening to kill her so he could bring another woman into the house. I sympathsized with her plight, but I had no authority to offer her shelter that day because my superior was absent – but was due back the following day – so I promised to do something for this woman the next day.

Later that night, I suddenly heard shrieks of blood-curdling agony. I ran outside to see what was going on and came upon a horrifying sight: a woman totally engulfed in flames. When she saw me, she ran toward me screaming "Save me! Save me!" I suddenly recognized her as the very woman who had come to me earlier that afternoon. Her husband, in a drunken rage, had poured kerosene on her, and set her on fire – a practice that is tragically not uncommon in India. I grabbed a blanket and smothered the fire as she fell to the ground. With the help of some onlookers, we took her to a hospital. But she died that night of 90% burns, along with her seven month old foetus.

I was absolutely devastated. I could not forgive myself for not having helped her earlier that day. I had been brought up in a secure family environment in Kerala, and I had no idea that one night could make such a difference in the life of a woman. I wanted to run away from the world, and its cruelty and wickedness. My friends, and especially a priest friend named Father D'Sa, dissuaded me from becoming a recluse, and helped me choose to do something about it instead of running away.

It was then that I decided I had to create a home for abused and traumatized women – a place where they could feel secure, cared for, and loved, irrespective of their religion, caste, or social status. It took some years to convince others, to raise the seed funds, but finally Maher opened the doors of its first house in 1997. Two women came that very first night.

Could you briefly describe the Maher project today?

Today Maher is a community of approximately 180 women and 600 children, plus a team of 15 social workers, 8 administrative staff, and a board of 9 trustees. About half the women are 'housemothers' who care for the children and the incoming battered women. Every housemother was once a battered woman herself who sought refuge at Maher, and her healing is enhanced by the responsibility of giving love and care to the other women and children. About half the children and women live at our main facility in Vadhu village, about 32 km east of Pune. The remainder live in nine satellite 'mini-Maher' homes that have been established in surrounding villages. Each satellite community has 2 housemothers, who feed and care for about 20 children.

What are the principal features of the Maher community?

Our outreach programs help to lessen the causes of poverty, neglect, violence, and superstition, while spreading the benefits...

Maher began as a project for battered women and their children, but it has grown into a thriving community that moves beyond the relief of social symptoms to address the causes of these problems in Indian communities. We do this in five principal ways: First, we create a caring and nurturing environment for the women and children at Maher, where everyone is loved and respected. Second, Maher is an interfaith, intercaste project. This means we are committed to transcending all religious and caste barriers, which are some of the most destructive forces in Indian society. A woman in distress who takes refuge at Maher is welcomed here with open arms--regardless of her religion, caste, or socio-economic status, which to our knowledge is very rare in India. Third, we provide a range of essential skills, education, counseling, resources, and other services – not only to our own community, but also to many surrounding communities in the region. Our outreach programs help to lessen the causes of poverty, neglect, violence, and superstition, while spreading the benefits of Maher's work. Fourth, we have created a home for women found on the street who have been completely abandoned by society, and who are often mentally disturbed. We give these women a simple home, loving community, and dignified life. We also reach out to help the dalits ('untouchables') and 'tribals' in the area, who receive no assistance from the government. Finally, we have deep respect for nature and the interdependence of all life, which is reflected in everything we do. At Maher we have implemented many ecological and sustainable practices to minimize wasteful consumption and preserve the natural environment.

How does the 'interfaith' aspect of Maher work in practical terms?

Each person must choose for themselves which particular faith is their heart's true calling. At Maher, we honor all religions, and all scriptures. We celebrate several religious holidays, such as Divali (Hindu), Christmas (Christian), Id (Muslim), and Budh Purnima (Buddhist). Our prayer sanctuary has copies of major scriptures like the Bhagavad Gita, the Bible, the Qur'an, the Dhammapada, with artwork depicting each of these traditions. But we never promote one religion over the others in our community. Maher thrives by living in accordance with the universal divine Spirit that is at the very foundation of every major religion.

Maher was founded on a deep faith in two things; God, and the inherent goodness of people. That goodness is a reflection of the Divine in each of us, and it doesn't matter which faith we belong to. For example, I am a Catholic, which means that Jesus is my personal guru. But I rarely speak about this in public, because it is my personal relationship with God. Someone else has her personal relationship with the Divine. Each person must be free to worship whomever and however they choose. Maher's deputy director, Hirabegum Mulla, is a Muslim. The chair of our Board is a prominent teacher of Vipassana meditation in the Buddhist tradition. Our other Board members are Hindu, Christian, Buddhist, and Muslim. We have many Hindus and Muslims among our staff and residents. The unifying thread is that we all honor the Divine. On our own, we pray or meditate in our different ways, but in Maher community gatherings, we don't pray to any particular deity, but to 'God' only, whatever that might mean to each of us.

How does the intercaste commitment work in practice?

Maher does not recognize caste distinctions. In all our community activities, people of all castes are mixed together, which breaks traditional Indian social taboos. We have community dinners where people of the highest and lowest caste sit side by side eating together, which they never do outside. We reach out to local communities of dalits (untouchables) and indigenous 'tribals', who live in extreme poverty. They are actually below the lowest rung in the caste system, and are not even recognized by the Indian government, so they get no services whatsoever. Maher provides assistance to these people, such as water wells and pumps, solar cookers, and basic supplies. We also built

> *Each person must choose for themselves which particular faith is their heart's true calling. At Maher, we honor all religions, and all scriptures*

schools and day-care centers for their children, who otherwise would have no educational opportunities.

Our intercaste commitment has led to wonderful learnings. For example, we were receiving a shipment of 15 kilos of fresh vegetables every week from an anonymous donor. It took us a long time to find out who this donor was, and when we finally went to his house to thank him, we were shocked to discover that this man was exceedingly poor, of the lowest caste. He lived in a tiny one-room hut with his wife and four daughters. He eked out a living as a vegetable vendor, and managed to set aside 15 kilos for Maher every week. And he adamantly refused to accept any gift from us by way of thanks – he said that his family had plenty to eat, and he simply wanted to help other women and children. So a few months later at the inauguration of our new building, we invited him to be our Chief Guest and keynote speaker. This meant that we had high caste Brahmins and prominent businessmen who had funded the building sitting there in the audience, listening to the keynote talk given by this humble vegetable vendor. He had given in such purity of spirit, from such meagre means, without ever wanting even to be recognized for it. It was an inspiration to all of us.

How are battered women cared for and rehabilitated at Maher?

Women admitted to Maher are first given the necessary medical treatment and psychological counseling. Once they are stabilized and become adjusted, we work with each one on an individual basis, depending on her needs and circumstances. Most of them would like to reconcile with their husbands or families, and we help them with that, through counseling, legal support, etc. In cases where this is impossible, or they don't wish to go back to their husbands, we assist them in filing for divorce, or support them to develop the skills or training they need to support themselves and their children. It can take a long time in some cases. We also train the women in vocational skills such as handicrafts and tailoring, which are an integral part of life at Maher and serve as a source of income when they leave. Maher sells the products to provide a modest source of revenue for the women.

How does Maher serve the surrounding communities?

We have created 300 'self-help groups' in more than 85 villages in the area. Each group has 20 people that make small donations of money into a community pool. Any member can then request a loan at a nominal interest rate, and the proceeds go back into the community pool. Self-Help Groups eliminate dependence on unscrupulous loan sharks, who charge interest rates from 60 to 100 percent. The Self-Help Groups also serve as important avenues for raising awareness in rural villages about health, hygiene, domestic violence, alcoholism, and superstitious beliefs. Most self-help groups are strictly for women only, but we also have 10 self-help groups for men, because it is important to work with the entire communities, and not only with women.

In the villages surrounding Maher, competent medical help is often not available. As resources permit, we send out doctors and nurses to surrounding villages to provide free check-ups and medical care. Maher children attend nearby primary and secondary schools, but many women of Maher have dropped out or been deprived of education. Maher has created facility for National Open School and Adult Literacy classes to give basic skills and training to these women and other young adults.

What are the major ecological aspects of Maher?

Many environmentally friendly technologies are employed at Maher to minimize pollution and the consumption of non-renewable resources. Solar thermal collectors on the rooftops generate most of our hot water. We also have solar cookers and even solar powered lamps that save costly electricity at night. Food and vegetable wastes are composted using vermiculture systems, which utilize worms to process organic waste into highly effective fertilizer. The process requires little maintenance and produces no offensive odors. We minimize use of plastics, replacing them with recycled forms of paper and cloth bags whenever possible.

No chemical fertilizers or pesticides are used within the premises of Maher. All agriculture and gardening on our land is done with organic farming techniques, and we produce our own biogas for cooking. We adopted a total waste management system, in which solid wastes are used as manure, and liquid wastes are carried by separate pipeline to irrigate the garden. Children are encouraged to take up gardening and are allotted plots to maintain, and plants to care for. We teach them about the fundamental interdependence of all forms of life, and encourage them to develop a deep respect and love for nature. We strive for a high degree of consciousness in relation to ecology and the natural environment. On a larger scale, there are tree plantation drives in our region that Maher participates in, as well as occasional watershed projects for areas where water shortage is acute.

We strive for a high degree of consciousness in relation to ecology and the natural environment.

How does Maher succeed so well in the larger Indian society, which is well known for deeply entrenched patriarchy and corruption?

As I said, at Maher we trust the inherent goodness of people. Of course we have our share of problems. But we have never paid a bribe at Maher, and we never will. Sometimes this has caused delays for us, in getting a building permit, or getting the electric utility connected – sometimes upwards of two years or more. Once when we kept refusing to pay a bribe for a permit, some officials came out to visit us. We showed them around and were very hospitable to them, just like any other visitors. They liked our project very much. When they drew me aside and tried to collect their bribe in hushed tones, I asked them how much money they were asking for. They quoted a figure, and I told them to come with me. I led them into our main hall where many women were busy producing crafts and cooking, and the children were focused in their study circles. I quietly whispered to the officials that for that amount of money, we would have to put out at least 4 women and 6 children onto the street. I asked them to please go around the room and pick out which individuals should be put out. They suddenly became very quiet, then turned and left. Not long afterwards we got our permit.

What is your dream for the future of Maher?

I would love to see many more people come forward and truly commit themselves to the vital work of building communities of healing and love. It is such fulfilling work – you can't imagine! And the need is so great! Not only in India, but everywhere. You know, one of my favorite lines from the Bible is "The harvest is plentiful, but the laborers are few." Sadly, this is so true. Here in India, poor as this nation is, our scarcest resource for expanding Maher is not money; it is finding deeply committed people. The money will always come, but where are the truly committed souls? So my prayer for the future is that the fire of real Love will ignite more people's hearts, and inspire them to join this vital work. It can be done anywhere, and is needed everywhere.

Sister Lucy Kurien is a Catholic nun born in Kerala, South India. She founded the Maher interfaith refuge in 1997 for battered and downtrodden women and children outside Pune, India. Maher honors all religions and repudiates the caste system, and has provided rehabilitation to 2,000 battered women. Sr. Lucy's work and the remarkable healing stories of Maher are documented in the book *Women Healing Women* by William Keepin and Cynthia Brix (Hohm Press, 2009). Website: www.maherashram.org

Rashmi Mayur, director of former International Institute for Sustainable Future in Mumbai, and a leading spokesperson for the South, wrote about the health problems in the global South seen from a holistic point of view for the Johannesburg UN Sustainability Conference in 2002. Whilst some of his numbers have escalated, the fundamental picture he painted is unchanged.

Health in the Global South

Rashmi Mayur

The lack of these resources – water, food, and fertile land – gravely threatens the health of the people in developing countries. Water-borne diseases such as cholera and typhoid, which have been virtually eliminated in the North, are still rampant in the South. Half the world's people lack water-based sewer and sanitation systems and a billion people lack clean drinking water. In Manila, 40 per cent of the residents live in slums. Many build huts set up on poles with streams of sewage running underneath. Children play in the sewage, then go home and eat – if the family has managed to collect any food that day – without washing their hands because there is no water (and also as spoons and forks are not part of their cultures.) This scenario is repeated in various forms throughout the South. Consequently, 3.4 million people – most of them children – die each year from water-borne diseases. With education and basic sanitation, most of these deaths would never occur. But water-borne diseases are not the greatest health threats facing the developing world. The three greatest health threats are posed by malaria, tuberculosis, and AIDS.

According to the WHO, two billion people – a third of the world's population – show signs of latent tuberculosis infection. Tuberculosis kills 1.5 million people every year and, at the rate it is spreading, could kill 100 million people by the year 2050. The increasing rate of infection is due largely to the fact that people infected with HIV are more prone to TB, but there are new strains that are more virulent and drug-resistant than the ones we have known in the past.

On an average, there are 300 to 500 million cases of malaria every year in the world and between 1.5 and 2.7 million people die as a result – at least 80 percent of them in sub-Saharan Africa, where about 52 percent of the population carry malaria parasites in their blood. In Africa, 2,800 children die every day of malaria; in Brazil, the disease kills more people than AIDS and cholera combined. As we count all of the disease's costs, from medical care to lost productivity, we see that malaria takes almost $2 billion annually out of Africa's economies – an amount that is desperately needed for other things."

The situation is becoming worse: the anopheles mosquitoes that transmit malaria are increasingly becoming resistant to pesticides and malaria parasites are more and more resistant to the usual treatment drugs. The disease also will spread to new areas as global warming makes more geographic areas hospitable to the mosquito.

Developing new drugs is important, but we also must not ignore local medicines and the plant-based remedies that evolved through history. They may have an equally strong part to play in controlling the disease.

At the same time, we must eliminate the mosquitoes that transmit the malaria parasite. As a priority, national and global health organizations must identify the places where water stagnates and mosquitoes breed, and implement programs to eliminate these breeding grounds. Natural predators that eat mosquito larvae can be introduced to reduce the problem without the use of chemical pesticides and their harm intended or not, to human, animal, and ecological health.

Research, eventually leading to immunization, is where the major breakthroughs lie. However, the additional research needed is not done in the developing world; and, in the North, most of the resources needed for research are given to AIDS and to lifestyle related illnesses of the developed world, such as cancer and heart disease.

Of course, the focus of research on the treatment of AIDS is necessary and welcome. AIDS has become the fastest-growing health threat in the South and, after water-borne diseases, promises to become the greatest killer. Of the world's 40 million current cases of AIDS, 90 percent are in the South. As many as 28 million cases are in Africa, and by the end of this decade there may well be 20 million orphans in Africa because of the disease. As it is, 5,000 Africans die every day from AIDS. But the disease also has spread to China and India, and throughout the developing world.

The world needs $10 billion annually to manage the AIDS epidemic. The South does not have enough resources to deal with the education, treatment, medical facilities, and everything else, to combat this plague. This summit must begin to formulate a global approach to eradicating AIDS. Private foundations, such as the Gates Foundation in the U.S., are beginning to address the need. But these efforts must be coordinated under an international plan – and, of course, the efforts also must be vastly, indeed exponentially, increased.

Developing new drugs is important, but we also must not ignore local medicines and the plant-based remedies that evolved through history.

Indian-born Rashmi Mayur, Ph.D., was an environmental scientist, author, teacher, radio host, consultant to the U.N., organizer of the last Earth Summit in Johannesburg, South Africa. He died in 2004. Dr. Rashmi Mayur had devoted his life to educating the world's people on the state of the planet and of 'building a better earth'.

Dr. Cornelia Featherstone, who lives at Findhorn ecovillage and works as a general medical practitioner, gives an outline of what a wholesome, balanced and healthy lifestyle means her:

A Healthy Lifestyle

Dr. Cornelia Featherstone

Connecting with Spirit

Early morning meditation in the sanctuary allows me to connect with Spirit with my own small voice within that tells me that God is there in all for me to behold, that the intelligence of nature and other manifestations of consciousness are accessible to me – I just have to ask. Or I can go to Taize singing in the Nature Sanctuary to raise my voice in joy and devotion. Later in the day I can join group meditations or use the sanctuaries, special nature spots or wherever I am at that moment to re-connect and practice mindfulness, compassion or contemplation.

Exercise

A walk to the beach, some of us swim in the bracing Moray Firth every day from May through to November(!), or joining one of the exercise classes on offer. There is much choice: from Yoga, Tai Chi, to dancing (modern, Five Rhythms, sacred dance or belly dancing), to aerobics – at various times throughout the day. Some are classes, others are groups of friends getting together to support each other and have fun.

Food

A wholesome organic breakfast in a peaceful setting with my family, friends or by myself gives me sustenance to go into the day. Much of my food comes from the gardens and Earthshare (our Community Supported Agriculture project) – it is organic, local, in season, grown with love and with no food miles that cost the Earth. What the gardens or Earthshare cannot provide I find in the Phoenix shop – our community store which offers everything from the whole range of food to herbs, remedies, body care products to arts and crafts and books. My purchases support a local business that gives employment and brings wealth to the collective.

Work

My work inspires and fulfils me – I can express my care for my fellow human beings and for the Earth in a constructive way. I have the option of sharing about my inspiration, my concerns, visions or questions either in the work department or when meeting others. I know that my contribution is only one of many that make the whole community what it is and contribute to its work in the world.

Leisure

Arts, crafts and culture create a rich tapestry of joy, colour, and social networking. We are blessed with several very active arts centres – the Universal Hall for the performing arts, the Findhorn Pottery, the Arts Centre dedicated to create beauty, the Weaving Studio. There is a crafts group supporting a regular crafts fair. Many different opportunities to make music, in different choirs, ensembles and bands or just ad hoc when taking part in 'community sharings' (evenings or mixed performances in the Hall) or in the 'Open Mike' on a Sunday night, when we can let the performer in us off the lead. Many community members use 'spare time' for volunteer work. In our money driven society it is healing to the individual to be giving freely without counting the return. It can be for the joy of doing things with others, or for the sake of the task at hand – serving people or the environment, or just to experience the joy of giving generously, willingly – as an expression of service or abundance.

Governance

In the ongoing process of change, health is not a static condition. There are many things we can do to enhance our present state of health.

The empowerment gained from having a say in the wider context of our lives is an integral aspect of our community. Sometimes we may groan about the number of meetings and presentations to attend but they offer us a choice to be involved, to have our voice heard, to shape the life of the ecovillage. In the ongoing process of change, health is not a static condition. There are many things we can do to enhance our present state of health. To bring positive change it is important to identify the small, manageable next steps we can realistically take to improve our health, be that a change in diet, activity levels or relationship patterns. The next step is to make a commitment to that change and establish a support and review structure that will allow it to become anchored in the daily routine. Community support is essential for this – not only the infrastructure that offers a smorgasbord of opportunities but also the social support to sustain the change. The community offers a wide range of alternative medicine that can support the quest for improved life style and better health.

Dr. Cornelia Featherstone is a physician in the National Health Service and conducts research into the interface between complementary medicine and primary care. She lives in the Findhorn community, and founded the Findhorn Bay Holistic Health Centre in 1990 and HealthWorks, the Forres Centre for Holistic Health Care, in 1994. She has organised a series of international conferences, including Health Care for the 21st Century and The Medical Marriage. She is author of several scientific papers and is co-author of *Medical Marriage – the New Partnership between Orthodox and Complementary Medicine*.

Sabine Lichtenfels writes prayers for a new revolutionary spirituality that aims to support the transformation of both people and society. Here she offers wise counsel to help us create harmonious and nurturing relationships with our children.

The Dream of the Children

Sabine Lichtenfels

Give your children the possibility to connect with their dream.

You will no longer think: these are my children, they belong to me. All children are the children of Creation, they come from ME and return to ME. They come with their own plan for life.

It is possible that you will find in your children your cosmic companions and teachers.

Do not try to realize your own unfulfilled longings through your children. Realize them yourself.

Listen to the language of your children and sense their dream; this will awaken your own dream. Once you begin to see and understand their dream, children are a blessing for your own awakening.

Support them, so that their wisdom can come to them. They need your love, your strictness, your clarity, and your truth.

They especially need you to step out of relationship, so that they can remain free for their relationship with the world.

Connect with the dream of your children. This connection brings about a growth of power within you.

Protect your children by giving them the freedom that is theirs. Give them enough time alone, so that they do not lose their connection. Protect them from the ping pong of relationships. See to it that they remain connected to the source from which they came – even in your presence. It will never be as easy as during these first years of life to learn to see and understand the spirit of plants, animals and angels.

Protect this sacred space of the children by becoming aware of this connection with Creation. Do not disturb them if they happen to be connected with Creation as they dream awake. Instead, use this for your own reconnection. Children have their own angels, their guides, and companions. Accompany them on this journey of awareness, meditative stillness, and communication. Become their companions, too, and help them find orientation.

Do not be alarmed if they temporarily reveal themselves as revolting little monsters. They are briefly repeating the entire process of human evolution and history. Give them clear orientation here, too, based on your own power of stillness.

Give your children the opportunity to find and accept their cosmic being.

Ya Azim.

Listen to the language of your children and sense their dream; this will awaken your own dream.

From Sources of Love and Peace, *Sabine Lichtenfels, Meiga, ISBN 978 3 927266 11 7*

Sabine Lichtenfels is a theologian and the co-founder of Tamera community in Portugal. She has been engaged in community-based research for 30 years and co-founded innovative communities for 'Peace Research' and 'Healing Biotopes'. Sabine has led numerous pilgrimages for peace in Israel, Palestine, India, and Colombia. She is the author of several books including *Grace: Pilgrimage for a Future Without War*.

MODULE 5
Socially Engaged Spirituality

Contents

InterSpirituality:
Bridging the Religious & Spiritual Traditions of the World

The Spiritual Imperative

Guidelines for Socially Engaged Spirituality

Silence and the Sacred:
Interviews with Craig Gibsone and Robin Alfred

Spirituality in Damanhur

A Brief Snapshot of Auroville, India

Plum Village:
A Spiritual Perspective on Community

The Awakening of the World, the Village,
the Nation and the World:
The Sarvodaya Vision for our Global Future

A Hopi Elder Speaks

It is possible that the next Buddha will not take the form of an individual. The next Buddha may take the form of a community – a community practicing understanding and loving kindness, a community practicing mindful living. This may be the most important thing we can do for the survival of the Earth.

Thich Nhat Hanh

We begin this final section with one of the most pressing challenges of our time: the need for bridging among humanity's spiritual traditions, and to recognise that their differences are dwarfed by their deep commonalities. Nothing is more important than bringing more love into the world, which is the true goal of all religions.

InterSpirituality: Bridging the Religious and Spiritual Traditions of the World

William Keepin

"I am a Muslim, and a Hindu, and a Christian, and a Jew – and so are all of you!" Thus proclaimed Mahatma Gandhi in a moment of inspired exasperation, as his advisors were pressuring him not to meet with Muslim leaders during the struggle for India's independence. Gandhi was pointing to the universal truth of the heart, which is the birthright of every human being and is found in every major religious and spiritual tradition.

A thousand years from now, humanity will look back on our period in history as the pivotal time when this crucial insight of Gandhi's finally began to emerge on a broad scale. The hundred year period from about 1950 to 2050 will stand out as one of the most crucial turning points in human history; the time when the urgent work of healing and reconciliation across the religious divisions of the world finally took place. It could not have begun earlier because this transformation required the improved systems of transport and communication developed in the late 20th century, especially the airplane and the internet. Nor could it have waited until later, because the religious divisions afflicting the human family by the early 21st century were so deeply entrenched, and the available weapons of mass destruction so powerful, that the world was teetering on a global war of catastrophic proportions.

Never in human history has it been more important to bridge the diversity of world religions. Each religion offers a unique doorway to the spiritual truth that dwells within and beyond all beings. Each of these magnificent doors is exquisitely crafted in itself, and our basic dilemma is that we get trapped in worshipping the doors themselves and fighting over which door is most beautiful, rather than actually going through the doorway and merging with the one spiritual reality.[1]

The human heart is crying out for new bridges of love, healing, and reconciliation across differences – not only religious differences, but also racial, gender, caste, and social and class disparities as reflected in the Occupy Wall Street movement. Enlightened religious and political leaders are calling for greater harmony and collaboration across faith traditions. "The stakes are higher than ever," says the Dalai Lama, "not only for the survival of our species but also for the very planet itself and the myriad other creatures who share our home."[2]

In answer to this call, the world religions are slowly beginning to come together in new ways, despite their vast and rich differences. Teachings of compassion and deep respect for others is universal in all religions, and the 'golden rule' is found in every tradition.[3] Once, when a group of esteemed Rabbis were confronted with a challenge to teach the entire Torah in a few seconds, most scoffed in disdain at this absurd effrontery. Yet the great Rabbi Hillel suddenly broke the silence, "Love your neighbor as yourself. The rest is all commentary, go study."[4] Of course the human family admittedly falls far short of manifesting this profound love on a broad practical scale. The reality across the globe today still includes tragic religious strife and conflicts that continue to threaten global peace and stability, with the killing on all sides carried out in the name of God. Yet in the end, the only realistic path forward is for the entire human community to live as one family and one species in harmony with millions of other species on this planet.

Interfaith and Interspirituality

Spirituality today is rapidly changing around the world. Believers in traditional faith communities are stretching beyond their own traditions for new insights and practices. Membership in many traditional churches is waning, especially in Western countries. Pope Benedict XVI has lamented the weakening churches in Europe, Australia and the USA. "There's no longer evidence for a need of God, even less of Christ," he told Italian priests in 2005, "the so-called traditional churches look like they are dying."[5] A recent cover story in *Newsweek* magazine proclaims "Forget the Church, Follow Jesus."[6] Meanwhile non-traditional spiritual movements are growing rapidly. Independent spiritual pioneers such as Eckhart Tolle, Marianne Williamson, and Neale Donald Walsh are attracting large numbers of followers. Numerous spiritual schools have emerged that focus on specific esoteric teachings or certain 'New Age' writings such as *A Course in Miracles*. The recent YouTube video, 'Jesus is greater than religion', received over 18 million views in less than one month on the internet. Even leading evangelical Christian pastors such as Rob Bell are beginning to question openly the narrow orthodoxy of their traditions. In 2010 the United Nations declared the first week of February to be the annual 'World Interfaith Harmony Week'.

Although the religions of the world exhibit great differences, these outer disparities are dwarfed by the vast and universal truths that dwell in common at the mystical core of every religion.

The term 'Interspirituality' was coined by the late monk Wayne Teasdale, and refers to the common heritage of humanity's spiritual wisdom, including the sharing of resources, practices, and dialogues across the traditions.[7] Although the religions of the world exhibit great differences, these outer disparities are dwarfed by the vast and universal truths that dwell in common at the mystical core of every religion. Reverend Cynthia Brix provides a helpful distinction between interfaith and Interspirituality. "At their core, religions are one, and this oneness defines a universal or interfaith mysticism. Interspirituality joins the teachings from two or more traditions, through study and practices, to create an actual new spiritual path. The term 'interspirituality' also serves to open religion up more broadly for people who are deeply spiritual, but don't necessarily identify with any of the world traditions, and provides a place for them to sit in the circle."[8]

Interspiritual and interfaith organizations are expanding rapidly, building bridges and dissolving rigid barriers across the world's religious traditions. In 1945 there were three interfaith organizations in the world. Today there are over 2,000 interfaith organizations in the United States alone, half of which have appeared since 2003.[9] The World Alliance of Interfaith Clergy lists ten interfaith seminaries, including four in New York alone, and one each in Canada, UK, Mexico, and one forming in the Netherlands. Taken together, these seminaries collectively graduate approximately 250 interfaith ministers each year, who are forging a new vocation of interfaith ministry.[10]

Interspirituality is also growing in several other directions. Seekers within existing faith traditions are expanding their spiritual identities to include two or more religious traditions. Christians for example are training intensively in Buddhist meditation or Hindu Vedanta, or joining Sufi communities. This phenomenon of 'multiple religious belonging' is becoming more widespread and gradually more accepted.[11] In truth, such mixing of traditions is nothing new, and cross-fertilization between religions has a rich history. Zen is a blend of Buddhism and Taoism. Sikhism emerged from the reconciliation of Hinduism and Islam. Place the Jewish Shema (love of God) alongside Hillel's teaching cited above, and you get the twin commandments of Christianity. The Bahai faith upholds the unifying truth of nine world religions.

"Love is the answer. You are the question." Thus reads the byline on the 'spiritual but not religious' website, representing a growing community of seekers that includes between 50 million and 70 million people in the United States alone, according to organizer Steve Frazee. Other pioneering organizations include the Order of Universal Interfaith, founded to fulfil the dream of Wayne Teasdale to create an independent interspiritual fellowship that serves as an open ecclesiastical umbrella organization for many new and emerging spiritual initiatives. The Parliament of World Religions, founded in 1893 and resurrected a century later, now brings some 8,000 people of all religious faiths together every five years. The Global Peace Initiative of Women organizes international conferences to bring together senior religious leaders from multiple faith traditions, and to inject their collective wisdom into leading political spheres. Taken together, these multiple new threads are weaving a new spiritual fabric in society at large.

A unique breakthrough in the field of interspirituality is the Snowmass Conference founded by Father Thomas Keating, which brought seasoned religious leaders together from nine major world religions including: Buddhist, Hindu, Jewish, Islamic, Tibetan Buddhist, Native American, Russian Orthodox, Protestant, and Roman Catholic. Meeting over the past 30 years, this diverse group developed eight 'Points of Common Agreement' that form the basis for a kind of universal spirituality, although they did not name it as such.[12] Having established this remarkable common ground, and becoming good friends in the process, these leaders then proceeded cautiously to explore their religious differences. To their surprise and immense delight, they found that the widely divergent perspectives and practices within their various faith traditions became a source of deep insight and mutual inspiration. In the end they bonded even more over their differences than they had over their similarities.

The eight points of common agreement are the following:

1. The world religions bear witness to the experience of Ultimate Reality, to which they give various names: Brahman, Allah, Absolute, God, Great Spirit, etc.

2. Ultimate Reality cannot be limited by any name or concept.

3. Ultimate Reality is the ground of infinite potentiality and actualization.

4. Faith is opening, accepting, and responding to Ultimate Reality. Faith in this sense precedes every belief system.

5. The potential for human wholeness – or in other frames of reference, enlightenment, salvation, transformation, blessedness, nirvana – is present in every human person.

6. Ultimate Reality may be experienced not only through religious practices but also through nature, art, human relationships, and service of others.

7. As long as the human condition is experienced as separate from Ultimate Reality, it is subject to ignorance and illusion, weakness and suffering.

8. Disciplined practice is essential to the spiritual life; yet spiritual attainment is not the result of one's efforts, but the result of the experience of oneness with Ultimate Reality.

The Snowmass Conference demonstrates the essential oneness of all religions, and also reveals how the myriad differences among the world religions can be a source of mutual inspiration and illumination, rather than conflict. What the Snowmass Conference achieved on a small scale serves as a powerful beacon of light for what is possible in reconciliation among the world's religions on a larger scale.

The Interspiritual Path of Love

Love is a supreme and unifying principle in all the world's religions. Striking parallels exist, for example, in the foundational teachings of Christianity, Hinduism, Islam, and Judaism – which taken together point to the possibility of a universal path of love across the faith traditions. These deep parallels are remarkable and inspiring, yet sadly often go unnoticed, because excessive emphasis is placed today on the differences between religions – differences that are generally far less profound than the similarities.

In exploring these parallels, it is important to be mindful of key differences so we don't superficially fuse distinct traditions or theologies that do not belong together. Nevertheless, it is also important to allow for the possibility of a universal truth or foundation which all traditions reveal, but which may appear differently across the faith traditions because it is rendered through dissimilar theological concepts and frameworks.[13] Recent developments in the interspiritual path of love, summarized below with a few brief examples, are working through these issues and pointing the way toward a universal path of love.

Religious exclusivism and the struggle for supremacy among religions is contrary to the teachings of every tradition. Scriptures from each major religion include injunctions that call for respect of other religions.[14] Yet during religious conflicts, spiritual truths common to both warring parties are generally trampled underfoot. As the great Indian poet Kabir lamented in relation to conflicts between Hindus and Muslims, "O brethren! Allah and Rama are but different names for one and the same Being. The entire world is laboring under a great harmful delusion. One swears on the Veda, and another on the Qur'an. There is really no difference between these two paths. The two fight each other over a name, and perish. The true Reality neither knows!"[15]

Spiritual parallels between Hinduism and Islam are plentiful and deep, as reflected in the respective scriptures, the Bhagavad Gita and the Qur'an. The concept of the Godhead presented in these two scriptures is virtually identical, as is the injunction to worship only God, and none other. The parallel teachings are so numerous, and so strongly outweigh the differences between the two traditions, that scholar Pandit Sunderlal concludes "in regard to the basic duties of human life... and how to prosper in this life and attain salvation in the world to come, the Gita and the Qur'an hold but one and the same view."[16]

Terrorism and violence are against the teachings of every major faith

Religious exclusivism and the struggle for supremacy among religions is contrary to the teachings of every tradition.

tradition, and violence in all scriptures is admissible only as a last resort, and then only in self-defense. The 'support' for violence or war that is ascribed to both the Qur'an and the Bhagavad Gita can only be understood in the historical context when these scriptures were written. In both cases war was the last resort of a people facing total annihilation from an enemy bent on their destruction. Even Mahatma Gandhi, the great advocate of nonviolence, said that "where there is only a choice between cowardice and violence, I would advise violence... I would rather have India resort to arms in order to defend her honour than that she should, in a cowardly manner, become or remain a helpless witness to her own dishonour... But I believe that nonviolence is infinitely superior to violence." Both the Gita and Qur'an teach the same. Contrary to widespread popular belief, the notion of Jihad as 'Holy war' appears nowhere in the Qur'an. Jihad means 'effort' required to advance spiritual development and social harmony, and except when facing a mortal danger, Muslims are commanded in the Qur'an to repel evil with something better, so that their enemy will become an intimate friend.[18]

Deep spiritual parallels also hold between Christianity and Hinduism, where remarkable correspondences between the teachings and lives of Christ and Krishna are found but rarely highlighted. Both Jesus and Krishna are proclaimed as human Incarnations of the Divine. Jesus is one with God the Father, and Krishna is one with Purushottama, the Godhead. Each speaks in the first person with the authority and word of God.

Both Christ and Krishna proclaim themselves to be the exclusive pathway to God. Jesus says "I am the way, the truth, and the life. No one comes unto the Father except through me." (*John 14.6*). Krishna makes a very similar proclamation in the Bhagavad Gita , e.g. "I am the Way, the Supporter, the Lord. I am the Father of the Universe" (9.18, 17). Yet there is no contradiction here. How can this be? The 'I am' that Jesus speaks of is *ego eimi*, the name that God declared of Himself to Moses.[19] It is the eternal Divine voice that speaks through all prophets and saints in all cultures across the ages. This is why Jesus says "I am before Abraham was," (*John 8:58*), even though Abraham lived centuries before Jesus. Krishna makes virtually identical anachronistic statements in the Bhagavad Gita. In the one case, Christ-merged-with-God speaks in the first person through and as Jesus, and in the other case the Purushottama (Godhead) speaks through and as Krishna, who also uses the personal pronoun 'I'. As theologian Raimon Panikkar explains, "Christ is the Christian symbol for the whole of Reality" and other religions use different names for this same Reality.[20] Christ and Krishna each affirms, in his own time and manifestation, the unique pathway to oneness with the Divine.

Toward a Universal Path of Love

The examples presented above are but a small sample of many spiritual parallels to be found among Christianity, Hinduism, Judaism, and Islam.[21]

We need not concern ourselves here with why these parallels exist, or what their origin is, although these are certainly legitimate questions. The point here is that these spiritual parallels are readily found by anyone who delves deeply into the respective traditions and scriptures, rather than be satisfied with the superficial representation of religions presented in the media and certain academic and clerical formulations.

Different religions can be likened to the trees in a forest, which are completely alike and totally different at the same time. They are alike because they all have one trunk and multiple branches, they all sink roots into the same ground and draw water from the same invisible source, they all stretch vertically upward toward the one dazzling light above. Yet they are different because some have leaves, others have needles, some are deciduous, others are evergreens, and so on. Yet despite their many important and wonderful differences, trees are ultimately far more alike than they are different. So too with religions.

Important new bridges are being established here – between diverse religious traditions, between West and East and Middle East. These bridges span the enormous chasms of radically different cultures, traditions, and historical epochs. This gives new meaning to the (Christian) teaching that "Where two or more are gathered in My name, there I am also." A unique interspiritual cross-fertilization takes place when two or more religions come together, and this is the vital work so urgently needed today. As Mirabai Starr puts it, "By saying yes to the best of our own heritage and entering the holiest grounds of one another's faith traditions, we may be able to usher in an age of love."[22]

Taken together, the spiritual parallels outlined above point to a universal path of love that unites the human soul with the Infinite. This resplendent path of the heart is found within the religious traditions, yet is ultimately independent and goes beyond every religion. The path is open to all, through the gateway of the heart. The religious traditions did not create this path; rather, the fact of this path precipitated the creation of the religions. The purpose of the traditions is to lay out this path clearly, and to support the soul in taking up the journey, nothing else. The religious traditions themselves are but pointers to the path. The actual journey can only be taken by the aspiring soul.

"God is humanity's secret. Humanity is God's secret. This is the secret of secrets."[23] Ultimately we must each discover this secret ourselves, and take the plunge ourselves into 'the dark silence in which all lovers lose themselves' (Ruysbroeck).[24] For in that dark silence is the Real, the Beloved, the One without a name – and never can we rest until we merge in union with That.

A unique interspiritual cross-fertilization takes place when two or more religions come together, and this is the vital work so urgently needed today.

1   Gratitude to Tessa Maskell and Martin Cecil for suggesting this metaphor.
2   H.H. Dalai Lama, *Toward a True Kinship of Faiths*, Harmony, 2010, p.182.
3   For versions of the 'Golden Rule in 21 different world religions', see www.religioustolerance.org/reciproc.htm.
4   http://gtorah.com/category/sources/hillel-shamai/
5   Quoted in 'Religion takes a back seat in Western Europe', *USA Today*, August 10, 2005.
6   Andrew Sullivan, 'The Forgotten Jesus', *Newsweek*, April 9, 2012, pp.26-31.
7   Wayne Teasdale, *The Mystic Heart*, New World Library, 2001.
8   Cynthia Brix, 'Are You Interspiritual', *Integral Yoga*, Summer, 2011, pp.12-13.
9   Rev. Dr. Dan Rosemergy, National Interfaith Alliance, 'Building Interfaith Communities: The Challenge and the Vision', Big-I conference, Nashville, TN, 5 February 2012. Data for 1945 from Prof. Darrol Bryant, Center for Dialogue and Spirituality in the World's Religions, Renison University College of Waterloo, Ontario, Canada, presentation to Big-I conference, Nashville, TN, Feb. 5, 2012.
10 Rev. Philip Waldrop, Chair of Board, A World Alliance of Interfaith Clergy, private communication.
11 As one example, see the book by esteemed theologian Paul Knitter, *Without Buddha I Could Not Be A Christian*, Oneworld, 2009.
12 *The Common Heart: An Experience of Interreligious Dialogue*, ed. Netanel Miles-Yepez, Lantern Books, 2006.
13 For a deeper exploration and analysis of these parallels in Christianity, Hinduism, and Islam, see W. Keepin, 'The Interspiritual Path of Divine Love', presentation to Big-I Conference, Nashville, TN, 2012.
14 For examples of teachings from six major religions that encourage respect for other religions, see 'Light of Universal Spirit', Ch. 4 in W. Keepin and C. Brix, *Women Healing Women*, Hohm Press, 2009, p.48, n.1.
15 Pandit Sunderlal, *The Gita and the Qur'an*, Pilgrims Publishing, 2005, pp.37-38.
16 Pandit Sunderlal, *The Gita and the Qur'an*, Pilgrims Publishing, 2005, p.20.
17 R. K. Prabhu & U. R. Rao, editors; from section 'Between Cowardice and Violence', of the book *The Mind of Mahatma Gandhi*, Ahemadabad, India, revised edition, 1967.
18 *Qur'an* 41.34. See the enlightening discussion in Mackenzie, Falcon and Rahman, *Religion Gone Astray*, Skylight Paths, 2011, pp. 66-80.
19 Ehyeh in Hebrew, Exodus 3.14.
20 Raimon Panikkar, *Christophany: The Fullness of Man*, Orbis, 2009, pp.143-155.
21 For a more detailed exploration, see W. Keepin, 'The Interspiritual Path of Love', forthcoming, 2012.
22 Mirabai Starr, *God of Love*, Monkfish, 2012.
23 Sufi saying, quoted by Llewellyn Vaughan-Lee, Golden Sufi Center, Inverness, CA.
24 Jan van Ruysbroeck, *The Adornment of Spiritual Marriage*, Chapter 4, Kessinger Publishing, 2007.

Please see page vi for William Keepin's biography.

Satish Kumar persuavisely argues why spirituality has to underpin civilisation; in business, politics, as a force for good in social change, and in our personal lives to enable humanity to create a better world, not only for ourselves, but for the natural world as a whole.

The Spiritual Imperative

Satish Kumar

Matter and Spirit are two sides of the same coin. What we measure is matter; what we feel is spirit. Matter represents quantity; spirit is about quality. Spirit manifests itself through matter; matter comes to life through spirit. Spirit brings meaning to matter; matter gives form to spirit. Without spirit matter lacks life. We are human body and human spirit at the same time. A tree too has body and spirit; even rocks which appear to be dead contain their spirit. There is no dichotomy, no dualism, no separation between matter and spirit.

The problem is not matter but materialism. Similarly there is no problem with spirit, but spiritualism is problematic. The moment we encapsulate an idea or a thought into an 'ism' we lay the foundations of dualistic thought. The universe is universe, one song, one poem, one verse. It contains infinite forms which dance together in harmony, sing together in concert, balance each other in gravity, transform each other in evolution and yet the universe maintains its wholeness and its implicate order. Dark and light, above and below, left and right, words and meaning, matter and spirit complement each other, comfortable in mutual embrace. Where is the contradiction? Where is the conflict?

Life feeds life, matter feeds matter, spirit feeds spirit. Life feeds matter, matter feeds life and spirit feeds both matter and life. There is total reciprocity. This is the oriental worldview, an ancient worldview, a worldview found in the tribal traditions of pre-industrial cultures where nature and spirit, Earth and heaven, sun and moon are in eternal reciprocity and harmony.

Modern dualistic cultures see nature red in tooth and claw, the strongest and fittest surviving, the weak and meek disappearing, conflict and competition as the only true reality. From this worldview emerges the notion of a split between mind and matter. Once mind and matter are split then debate ensues as to whether mind is superior to matter or matter is superior to mind.

This worldview of split, rift, conflict, competition, separation and dualism has also given birth to the idea of separation between the human world and the natural world. Once that separation is established, humans consider themselves to be the superior species, engaged in controlling and manipulating nature for their use. In this view of the world, nature exists for human benefit, to be owned and possessed, and if nature is protected and conserved then the purpose is only for human benefit. The natural world - plants, animals, rivers, oceans, mountains and the skies - is denuded of spirit. If spirit exists at all, then it is limited to human spirit. But even that is doubtful. In this worldview humans too are considered to be nothing more than a formation of material, molecules, genes and elements. Mind is considered to be a function of the brain, and the brain is an organ in the head and no more.

Spirit in Business

This notion of spiritless existence can be described as materialism. All is matter; land, forests, food, water, labour, literature and art are commodities to be bought and sold in the marketplace - the world market, the stockmarket, the so-called free market. This is a market of competitive advantage, a cut-throat market, a market where survival of the fittest is the greatest imperative: the strong competing with the weak and winning the biggest share of the market for themselves. Monopolies are established in the name of free competition. Five supermarket chains control eighty per cent of food sold in the UK. Four or five giant multinational corporations, such as Monsanto and Cargill, control eighty per cent of international food trade. Small and family farms cannot compete with the big players and are forced to retreat. This is the world where spirit has been driven out. Business without spirit, trade without compassion, industry without ecology, finance without fairness, economics without equity can only bring the breakdown of society and destruction of the natural world. Only when spirit and business work together can humanity find coherent purpose.

Spirit in Politics

Just as materialism rules economics it also rules politics. Instead of seeing nations, regions and cultures of the world as one human community, the world is seen as a battlefield of nations competing with each other for power, influence and control over minds, markets and natural resources. One nation's interest is seen in opposition to the national interest of another. Indian national interest is opposed to Pakistani national interest. Palestinian national interest and Israeli national interest; American national interest and Iraqi national interest; Chechen national interest and Russian national interest, and so on ... the list is long. And so we have polarised politics: "If you are not with us you are against us," has become the dominant mind-set. And if you are not with us you are not only against us, you are part of the axis of evil.

There can be no democracy and freedom without compassion, reverence and respect for diversity, difference and pluralism.

This is politics denuded of spirit. What can we expect from such politics other than rivalry, strife, the arms race, terrorism and wars? Politicians speak of democracy and freedom but they pursue the path of hegemony and self-interest. How can a particular view of democracy and freedom suit the whole world? There can be no democracy and freedom without compassion, reverence and respect for diversity, difference and pluralism. Compassion, reverence and respect are spiritual qualities - but politics founded on materialism considers the values of the spirit to be woolly, flaky, utopian, idealistic, unrealistic and irrational. But where has the politics of power, control and self interest led us? The First World War, the Second World War, the cold war, the Vietnam war, the war in Kashmir, the war in Iraq, the attack on the Twin Towers of New York. Again the list is very long. Politics without spirituality has proved to be a grand failure and, therefore, it is time to bring politics and spirituality together again.

Spirit in Religion

Sometimes the words spirituality and religion are confused, but spirituality and religion are not the same thing. Politics should be free from the constraints of religion but should not be free of spiritual values. The word religion comes from the Latin root religion which means to bind together with the string of certain beliefs. A group of people come together, share a belief system, stick together and support each other. Thus religion binds you, whereas the root meaning of spirit is associated with breath, with air. We can all be free spirits and breathe freely. Spirituality transcends beliefs. The spirit moves, inspires, touches our hearts and refreshes our souls.

When a room has been left closed, doors and windows shut and curtains drawn, the air in the room becomes stale. When we enter the room after a few days we find it stuffy so we open the doors and windows to bring in fresh air. In the same way, when minds are closed for too long we need a radical avatar, a prophet, to open the windows so that our stuffy minds and stale thoughts are aired again. A Buddha, a Jesus, a Gandhi, a Mother Teresa, a Rumi, a Hildegard of Bingen appears and blows away the cobwebs of closed minds. Of course we don't need to wait for such prophets: we can be our own prophets, unlock our own hearts and minds and allow the fresh air of compassion, of generosity, of divinity, of sacredness to blow through our lives.

Religious groups and traditions have an important role to play. They initiate us into a discipline of thought and practice; they provide us with a framework; they offer us a sense of community, of solidarity, of support. A tender seedling needs a pot and a stick to support it in the early stages of its development, or even the enclosure of a nursery to protect it from frost and cold winds. But when it is strong enough it needs to be planted out in the open so that it is able to develop its own roots and become a fully mature tree. Likewise religious orders act as nurseries for seeking souls. But in the end we each have to establish our own roots and find divinity in our own way.

There are many good religions, many good philosophies and many good traditions. We should accept all of them and accept that different religious traditions meet the need of different people at different times, in different places and in different contexts. This spirit of generosity, inclusivity and recognition is a spiritual quality. Whenever religious orders lose this quality, they become no more than mere sects protecting their vested interests.

At present the institutionalised religions have fallen into this trap. For them the maintenance of institutions has become more important than helping their members to grow, to develop and to discover their own free spirit. When religious orders get caught in maintaining their properties and their reputation they lose their spirituality and then they, too, become like a business without spirit. As it is necessary to restore spirit in business and in politics we also need to restore spirit in religion. This may seem a strange proposition because the very *raison d'être* of every religion is to seek spirit and to establish universal love. The reality is otherwise. Religions have done much good but also they have done much harm, and we can see all around us that tensions between Christians, Muslims, Hindus and Jews are major causes of conflicts, wars and disharmony.

The rivalry among religions would cease if they realised that religious faiths are like rivers flowing into the same great ocean of spirituality. Even though the various rivers with their different names give nourishment to different regions and different peoples, they all provide the same quality of refreshment. There is no conflict among the rivers. Why then should there be conflict among the religions? Their theology or belief system may differ but the spirituality is the same. It is this spirituality which is paramount. Respect for a diversity of beliefs is a spiritual imperative.

Spirituality and Social Change

As business, politics and religious institutions need to return to their spiritual roots, so do the environmental and social justice movements need to embrace a spiritual dimension. At present most social change movements concentrate on negative campaigning. They present doom and gloom scenarios and become mirror images of the institutions they criticise.

The real impetus for ecological sustainability and social justice stems from ethical, aesthetic and spiritual visions. But this focus gets lost when campaigners get caught in false goals such as their desire to attract media attention or their need to gain more members for their organisations. These concerns become ends in themselves and the presentation of a holistic, inclusive and constructive vision is forgotten. Love of nature and the intrinsic value of all life, human as well as other than human, is the essential ground in which environmental and social justice movements need to be rooted. The basis of all campaigning is reverence for life, and this is a spiritual basis. There is no contradiction between pragmatic campaigning and a spiritual overview. Mahatma Gandhi's political programme was founded upon spiritual values. Martin Luther King's Civil Rights Movement was

The real impetus for ecological sustainability and social justice stems from ethical, aesthetic and spiritual visions.

rooted in a spiritual vision. Contemporary environmental and social justice movements also require that broad worldview rather than be limited to the science of ecology and the social sciences.

Spirituality and Science

Often it is believed that science and spirituality are like oil and water: they cannot mix. This is a mistaken notion. Science needs spirituality and spirituality needs science.

When science forsakes the restraints of moral, ethical and spiritual dimensions and strives to achieve everything that is achievable, experimenting with everything irrespective of consequences, then science leads to the technologies of nuclear weapons, genetic engineering, human and animal cloning and poisonous products which pollute soil, water and air. It is dangerous to give science carte blanche to dominate human minds and to subjugate the natural world. Contemporary science has acquired such status of superiority that it is presently commanding the total adherence of industry, business, education and politics. Some of its experiments have become so crude and cruel that it reaches beyond the constraints of civilisation. Ethical, moral and spiritual values are essential to moderate the power of science.

As science needs spirituality, spirituality also needs science. Without a certain amount of rational, analytical and intellectual skills spirituality can easily turn into sectarian and selfish pursuit.

As science needs spirituality, spirituality also needs science. Without a certain amount of rational, analytical and intellectual skills spirituality can easily turn into sectarian and selfish pursuit. I was a monk for nine years, pursuing my own purification and salvation. I saw the world as a trap and spirituality as a way of liberation from the world. Then I came across the writings of Mahatma Gandhi. He said that there is no dualism between the world and the spirit. Spirituality is not just for saints. It is not confined to monastic orders or caves in the mountains. Spirituality is in everyday life, from the growing of food to cooking, eating, washing up, sweeping the floor, building the house, making clothes and caring for neighbours. We must bring spirituality into all parts of our lives: into politics, into business, into agriculture and into education. And we must do so with a scientific approach.

That was such an inspiring insight that I decided to leave the monastic order and return to the world of everyday life.

Meeting Spiritual Needs

We human beings have our bodily needs and also our spiritual needs. Food, water, shelter, warmth, work, education and health are our essential needs. We need to engage in economic activities to fulfil these needs. But once these needs are met we need to find a sense of contentment and satisfaction in order to be happy and fulfilled. We need the wisdom to know when enough is enough. If we go on with economic activities even after our essential needs are met, then we become victims of greed and desires. Many of our social, political and environmental crises are crises of desire.

Those who profit from endless economic activities put enormous effort into persuading us that by having more material goods we shall be happy. But happiness does not come from material things alone; we also have social and spiritual needs: the need for community, for love, for friendship, for beauty, for art and music. We need to use our imagination and our creativity. We need the opportunity to make things with our own hands. We need time to be still and contemplate; we need spaces to appreciate and enjoy. These spiritual needs cannot be met by turning ourselves into consumers of goods provided by companies who make vast profits at the cost of the environment and ethics and at the expense of future generations. Materialism has become their new religion and they want everyone to be converted to it and become loyal members of their faith.

This religion of materialism is obviously unsustainable. If the six billion citizens of the world were to live the lifestyle of Western consumers and use the energy provided by fossil fuels we would need five planets, but we haven't got five planets: we have only one planet. Therefore, we need to invent a lifestyle of elegant simplicity where Earth's gifts are shared among all human beings fairly, without compromising the needs of the more than human world as well as of future generations. Such elegant simplicity is the way to discover spirituality. We embrace simplicity not only because the consumerist lifestyle is unfair, unjust and unsustainable but also because it is the cause of discontent, dissatisfaction, disharmony, depression, disease and division. Even if there were no problem of global warming, of resource shortage, of pollution and waste we would still need to choose a more simple lifestyle which is conducive to and congruent with spirituality, because a simple lifestyle, a lifestyle uncluttered with the burden of unnecessary possessions, is the lifestyle which can offer the opportunity to explore the universe of the imagination and to find boundless joy in that universe.

The Buddha was a prince; he possessed palaces, elephants, horses, land and treasures of gold and silver but he realised that all his wealth was holding him back, that wealth was keeping him chained to greed, desire, craving, pride, ego, fear and anger. The idea that wealth and power would make him happy was an illusion; joy through material possessions was a mirage. So he embraced a life of noble poverty which meant voluntary acceptance of limits. There was no population explosion at that time, the Buddha faced no shortage of raw materials or natural resources, there was no problem of global warming and yet he preferred the path of spirituality and simplicity because that was the way to meet the needs of the soul as well as the body.

Spirituality and Civilisation

My land, my house, my possessions, my power, my wealth are the cravings of small minds. Spirituality frees us from small mind and liberates us from the small I, the ego identity. Through spirituality we are able to open the doors of big mind and big heart where sharing, caring and compassion are the true realities. Life exists only through the gift of other lives: all life is interdependent. Existence is an intricately interconnected web of relationships. We share the breath of life and thus we are connected. Whether we are rich or poor, black or white, young or old, humans or animals, fish or fowl, trees or rocks, everything is sustained by the same air, the same sunshine, the same water, the same soil. There are no boundaries, no borders, no separation, no division, no duality; it is all the dance of eternal life where spirit and matter dance together. Day and night, Earth and heaven all dance together, and wherever there is dance, there is joy and beauty.

The religion of materialism and the culture of consumerism which have been promoted by Western civilisation have blocked the flow of joy and beauty. Once, Mahatma Gandhi was asked, "Mr. Gandhi, what do you think of Western civilisation?" He replied, "It would be a good idea." Yes, it would be a good idea because any society discarding spiritual values and fighting for material goods, going to war to control oil, producing nuclear weapons to maintain its political power cannot be called a civilisation. The modern, consumerist culture built on unfair, unjust and unsustainable economic institutions cannot be considered to be civilised. The true mark of civilisation is to maintain a balance between material progress and spiritual integrity. How can we consider ourselves to be civilised when we don't know how to live with each other in harmony and how to live on the Earth without destroying it? We have developed technologies to reach the moon but not the wisdom to live with our neighbours, nor mechanisms to share food and water with our fellow human beings. A civilisation without a spiritual foundation is no civilisation at all.

The way we treat animals is a clear example of our lack of civilisation. Cows, pigs and chickens live as prisoners in factory farms. Mice, monkeys and rabbits are treated as slaves as if they felt no pain; all for human greed and human arrogance. Western civilisation seems to believe that all life is expendable in the service of human desire. Racism, nationalism, sexism and ageism have been challenged and to some extent eradicated, but humanism still rules our minds. As a result we consider the human species to be superior to all other species. This humanism is a kind of speciesism. If we are to strive for civilisation we will have to change our philosophy, our worldview and our behaviour. We will have to enter into a new paradigm where all beings are interbeings, interdependent, interrelated and interspecies.

Spirituality Begins at Home

Where do we begin this spiritual revolution? We begin with ourselves. Self-transformation is the first step towards social, political and religious transformation. All transformations start at the bottom and move upwards to embrace the larger world. That is the law of the natural world. The great and mighty oak begins with the sowing of an acorn in the soil. After the seed is sown, for a few weeks or months no-one knows whether that acorn is living or dead or whether it will ever emerge into the world. But that unseen transformation under the earth's surface enables the acorn to emerge out of

The true mark of civilisation is to maintain a balance between material progress and spiritual integrity.

the soil as a tiny tender shoot. It is still small and insignificant but only from that insignificant beginning starts the process which eventually results in the mighty oak tree.

My mother used to say, "It is better to light a candle than curse the darkness, but before you can light other candles you need to light your own candle. Be your own light. Then you can offer yourself to help others. How can you make someone else happy if you yourself are not happy? But your happiness is born of your kindness to others."

So personal, social and political transformation go together because when we are free from fear and anxiety and at ease with ourselves then we are able to engage with the community around us and with society at large to bring about social and political changes to improve the lives of all. That selfless act of altruism in turn brings us a greater sense of fulfilment, satisfaction and happiness. Thus personal and political interact.

Three Practical Steps Towards Spirituality

Trust

So let us explore a few areas of spirituality. First and foremost among them is removal of fear and cultivation of trust. If we look deeply we will realise that many of our psychological difficulties stem from fear. A sense of insecurity, the ambition to be successful, the desire to prove ourselves, efforts to impress others, craving for power over others and to be in control, addiction to shopping, consuming and possessing, all are ultimately related to fear. This personal fear expands into social insecurity and political insecurity. So the first step towards spiritual renewal is to look at the phenomenon of fear in our lives and realise that much of this fear is aggravated by more fear. Fear breeds fear and fear is led by fear. We go to great lengths to build psychological and physical defences but they only increase our fear. Even when we have nuclear weapons to protect us we are not free from fear.

Moreover, history has proved that nuclear weapons are no defence and bring no security. The attack on the Twin Towers of the World Trade Center in New York proved that ultimately all defences are futile. The attackers can attack with a knife or a razor blade, so where is the justification for spending so much effort, time and resources in building nuclear warheads when they bring no defence and no security? The most powerful country in the world, the USA, is also the most insecure country in the world. Paradoxically the more defences we build the more insecure we are. Western societies seem to be obsessed with safety and security and go to great lengths to insure themselves against all eventualities. Such obsession has a paralysing effect.

The first step into the spiritual sphere is to understand fear and cultivate trust. Trust yourself. You are as good as you are. You embody the divine spark, the creative impulse, the power of imagination which will always be with you and will protect you. Trust others: they are in the same boat as you. They long for love as much as you do. Only in relationships with others will you blossom. You are because others are and others are because you are. We all exist, flourish, blossom and mature in this mutuality, reciprocity and unity. Give love and love will be reciprocated. Give fear and fear will be reciprocated. Sow one seed of thistle and you will get hundreds of thorny thistles. Sow one seed of camellia and you will get hundreds of camellia flowers. You will reap what you sow; this is the old wisdom. And yet we have not learnt it.

Then trust the process of the universe. The sun is there to nourish all life. Water is there to quench the thirst. The soil is there to grow food. Trees are there to bear fruit. The moment a baby is born the mother's breast is filled with milk. The process of the universe is embedded in the life-support system of mutuality. Hundreds of millions of species – lions, elephants, snakes, butterflies – all are fed, watered, sheltered and taken care of by the mysterious process of the universe; trust it. As Julian of Norwich said, "All shall be well, all manner of things shall be well."

The process of the universe is embedded in the life-support system of mutuality.

Participation

The second spiritual quality is participation. Participate in the magical process of life. Life is a miracle: we cannot explain it; nor can we know it in full, but we can actively and consciously participate in it without trying to control it, manipulate it and subjugate it.

Participation is easy and simple. We have been given two wonderful hands to cultivate the soil and grow our food. Working with the soil in the garden meets the need of the body as much as the need of the mind. Industrial farming has taken away our birthright to participate in the cultivation of food. Large-scale mechanised and industrialised farming is born of our desire to dominate. Small-scale, natural, local farming - still better, gardening - is a way of participating with the rhythms of the seasons. England should be gardened, not farmed. Animals should be freed from the prisons of factory farms. Growing food is one example of the principle of participation. Baking bread, cooking food, sharing the meal with family, friends and guests are as much spiritual activities as they are social and economic activities. The culture of fast food has deprived us of the fundamental activity of participation in the daily ritual and practice of physical and spiritual nourishment. It is wonderful that people all over Europe are inspired by the Italian movement of Slow Food. Slow Food is spiritual food. Fast food is fearful food.

Slowness is a spiritual quality. If we wish to restore our spirituality we have to slow down. Paradoxically only when we go slower can we go further. Doing less, consuming less, producing less will enable us to be more, to celebrate more, and to enjoy more. Time is what makes things perfect. Give yourself time to make things and give yourself time to rest. Take your time to do as well as to be. It is in the dance of doing and being that spirituality is to be found.

Once, the Emperor of Persia asked his Sufi Master, "Please advise me:

what should I be doing to renew my soul, revive my spirit, and refresh my mind so that I can be happy in myself and effective in my work?" The Sufi Master replied, "My Lord, sleep as long as you can!" The Emperor was surprised and amazed to hear this answer and said, "Sleep? I have little time to sleep. I have justice to perform, laws to enact, ambassadors to receive and armies to command. How can I sleep when I have so much to do?" The Sufi Master replied, "My Lord, the longer you sleep, the less you will oppress!" The Emperor was speechless: he saw the point of the Sufi sage. Even though the sage was blunt, he was right.

Western countries are in a similar position to the Emperor of Persia. The longer we work, the more we consume: we drive cars, fly in planes, burn electricity, go shopping and produce waste. The faster we do these activities, the more damage we inflict on the environment, on the poor and on our own peace of mind. So true participation is to live and work in harmony with ourselves, with our fellow human beings and with the natural world. Participation is not about speed and efficiency; rather it is about harmony, balance and appropriateness of action.

Gratitude

The third spiritual quality is a sense of gratitude. In our Western culture we complain about everything. If it is raining then we say, "Isn't it awful weather? So wet and cold!" When it is sunny we complain, "Isn't it hot? So hot!" The media are full of complaints and criticism. Debates in the parliament are mostly concentrated on the negative aspects of government policies. The opposition blames the government and the government complains about the opposition. The national culture of blaming and complaining permeates throughout, even in our family life and in our workplace. Because of the dominance of a culture of condemnation we learn to condemn ourselves too. "I am not good enough," is a widespread feeling. Whatever we do we don't appreciate it. We think we should be doing something different, something else, something better. Then whatever other people do we don't learn to appreciate it either. "I had a terrible childhood," we complain. "My school was awful," we reflect. "I'm never appreciated by my colleagues," we grumble, and this kind of criticism goes on and on.

In order to develop spiritually we need to balance our critical faculty with the faculty of appreciation and gratitude. We need to train ourselves to turn our minds to recognise the gifts we have received from our ancestors, our parents, our teachers, our colleagues and our society in general. We also need to express our thankfulness for the gifts of the Earth. What a wonderful Gaian system it is, that we are part of! It regulates climate, it organises the seasons and it provides abundance of nourishment, beauty and sensual pleasure to all creatures. When we are in awe and wonder at the workings of the sacred Earth we can feel nothing but blessed and grateful. When food is served we are filled with a sense of gratitude. We thank the cook and the gardener but also we thank the soil and the rain and the sunshine. We even

In order to develop spiritually we need to balance our critical faculty with the faculty of appreciation and gratitude.

express our gratitude to the earthworms who have been working day and night to keep the soil friable and fertile. However green a gardener's fingers are, without the worms there will be no food. So in praise we say, "Long live the worms," and further we join the poet Gerard Manley Hopkins and say, "Long live the wet and the wilderness yet." It is the beauty of the wild which feeds our soul while the fruit of the Earth feeds the body.

The generosity and unconditional love of the Earth for all its creatures is boundless. We plant one small seed of an apple in the ground. That tiny seed results in a tree within a few years and produces thousands and thousands of apples year after year. And all that from a tiny pip, sometimes self-sown. When in the autumn apples ripen with their fragrant, juicy, crisp flesh we eat to our hearts' content. The tree knows no discrimination; it asks no questions. Poor or rich, saint or sinner, fool or philosopher, wasp or bird, one and all can receive the fruit freely. What else can we feel for the tree but gratitude? And from our gratitude flows humility, as arrogance comes from complaining and criticism. When we are critical of nature we come to the conclusion that nature is not good enough: it is imperfect and unreliable. Nature needs our technology and engineering so we go to great lengths to improve on it, but we end up destroying it. With a sense of gratitude we go with the grain of nature, we work in harmony with it and we appreciate its miraculous qualities.

To summarise, the point I am making is that there is no dualism and separation between matter and spirit. Spirit is held within matter and matter within spirit but we have separated them and have made spirit a private matter and have allowed matter alone to dominate our public life. We need to heal this rift urgently. Without such healing, the material world, the Earth itself will continue to suffer catastrophic consequences, and spiritual insights and wisdom will continue to be seen as idealistic, esoteric and otherworldly practices totally irrelevant to our everyday existence.

When we are able to heal this rift we will be able to instil spirit in business, in commerce and in the economy. We will be able to create a politics which works for all. Our religions will not be divisive; on the contrary they will become a source of healing and resolving conflicts. The movement for environmental sustainability and social justice will inspire rather than agitate and, personally, human beings will be at ease with themselves and with the world around them. The marriage of matter and spirit, of business and spirit, of politics and spirit, of religion and spirit and of activism and spirit is the greatest union required in our time.

People are hungry for spiritual nourishment; this hunger cannot be satisfied by material means. Therefore, the great work we have in our hands is to create space and time for people to discover their spirituality as well as the spirituality of others.

It should not be necessary for me to make a case for spiritual space but because in the last few hundred years Western culture has been in denial of spirit and has been busy elevating the status of matter, our society and culture have lost their balance and wholeness. In order to restore this balance

I have emphasised the importance of spirit. In an ideal world people would recognise that spirit is always implicit in matter. Traditionally that is how it was. People took pilgrimages to holy mountains and sacred rivers; life was considered sacred and inviolable. We recognised the metaphysical dimension of trees. The speaking tree, the tree of knowledge and the tree of life express the implicit spiritual quality of the tree. Regaining this perennial wisdom is life's greatest imperative.

*This is the text of Satish Kumar's Schumacher Lecture given on 30th October 2004 in Bristol, UK (***www.schumacher.org.uk***) and was first published in* Resurgence Magazine *no 229.*

Born in India and starting life as a Jain monk, Satish Kumar is the editor of *Resurgence* Magazine (**www.resurgence.org**) which promotes an ecological and spiritually balanced way of living. He has also pioneered the Human Scale Education movement. He made an 8,000 mile peace pilgrimage on foot in the 1960s from India to America via Moscow, Paris, and London to deliver a humble packet of 'peace tea' from women in an Indian tea factory to the then leaders of the world's four nuclear powers. "Tell the leaders," the women said, "When you think you need to press the button, stop for a minute and have a fresh cup of tea."

Before we head out to create new communities or other forms of service, some key guidelines can help us to keep our actions pure and our relationships spiritually authentic. Fourteen principles of spiritual activism are presented briefly here that emerged from years of service work in the field.

Guidelines for Socially Engaged Spirituality

William Keepin

In facilitating inner work with social change leaders in Satyana Institute's programs over the years, a set of guidelines or principles emerged that support the practice of 'engaged spirituality'. These guidelines were developed for professionals and activists engaged in service work who wanted to bring greater love and spiritual wisdom into their daily professional lives. Attempting to formulate universal guidelines of this kind is of course fraught with peril, so the guidelines below are offered as a stimulus for inspiration, rather than as definitive principles. Taken together, they comprise a summary of insights that have proven helpful to socially engaged activists who grapple with how to integrate their 'inner' spiritual values into their 'outer' practical work.

Transformation

Transformation of motivation from anger/fear/despair to compassion/love/purpose. This is a vital challenge for today's social change leaders, particularly those who confront injustice in its various forms. This is not to deny the noble emotion of appropriate anger or outrage in the face of social injustice. Rather, it entails a crucial shift from fighting against evil to working for love, and the long-term results are very different, even if the outer activities appear virtually identical. Action follows Being, as the Sufi saying goes. Thus 'a positive future cannot emerge from the mind of anger and despair' (Dalai Lama).

Martin Luther King emphasized that we must purify our intentions before moving into direct action for social change. Otherwise the results of our work may actually undermine our noble purpose, in the name of advancing it. As Thomas Merton cautioned, "If we attempt to act and do things for others or for the world without deepening our own self-understanding, our own freedom, integrity and capacity to love, we will not have anything to give to others. We will communicate nothing but the contagion of our own obsessions, our aggressiveness, our ego-centered ambitions."

Non-attachment to Outcome

This is difficult to put into practice, yet to the extent that we are attached to the results of our work, we tend to rise and fall with our successes and failures – a sure path to burnout. Our task is to hold a clear intention, and let go of the outcome – recognizing that a larger wisdom is always operating. As Gandhi stressed, 'the victory is in the doing', not the results. Also, remain flexible in the face of changing circumstances: "Planning is invaluable, but plans are useless." (Churchill)

In Satyana's training programs, several social change leaders have reacted strongly to this principle. As one environmental lawyer stammered, "How can I possibly go into court and not be attached to the outcome? You bet I care who wins and who loses! If I am not attached to the outcome, I'll just get bulldozed! And when I lose, the Earth loses!" His exasperation underscores the poignant challenge of implementing these principles in the real world of political and social conflicts. Yet he kept coming back to our retreats, actively looking for ways to love his adversaries. He eventually came to see that non-attachment to outcome does not mean passive indifference to outcome. He also acknowledged that although it was difficult to love some of his adversaries, one way he could do so was to love them for creating the opportunity for him to become a passionate voice for truth and protection of the natural environment.

Non-attachment to outcome does not mean passive indifference to outcome.

Integrity is your Protection

If our work has integrity, this will tend to protect us from negative energy and circumstances. We can often sidestep negative energy from others by becoming 'transparent' to it, allowing it to pass through us with no adverse effects. This is a consciousness practice that might be called 'psychic aikido'.

Integrity in Means and Ends

A noble goal cannot be achieved utilizing ignoble means. Integrity in means cultivates integrity in the fruit of our work. Some participants in our trainings engaged regularly in political debates, testimony, and hearings. We suggested they apply the Tibetan tonglen practice for transmuting negative energy into compassion and love – right there in the hearing room. Those that experimented with this in earnest reported that it was very helpful in defusing charged psychological situations, and reducing tension in heated debates.

Don't Demonize Your Adversaries

It makes them more defensive and less receptive to your views. People respond to arrogance with their own arrogance, creating rigid polarization. Be a perpetual learner, and constantly challenge your own views.

The ideal is to constantly entertain alternative points of view, so that we move from certitude to perpetual inquiry. This is sometimes hard to do, because we often feel very certain about what we think we know, and the injustices we see. As John Stewart Mill observed, "In all forms of debate, both sides tend to be correct in what they affirm, and wrong in what they deny." Entering into an adversarial situation, we are acutely aware of the rightness of our own affirmations, but there is usually a kernel of truth in what is being affirmed by our opponents – however small. We need to be especially mindful about what we deny, because this is where our blind spots often lie.

You are Unique

Each of us must find and fulfil our true calling. "It is better to tread your own path, however humbly, than that of another, however successfully." (Bhagavad Gita) We each have a unique melody to contribute to the symphony of life. Discover yours, and sing it out with confidence, joy, and abandon – and let the harmony parts take care of themselves.

Love Thy Enemy

Or at least, have compassion for them. This is a vital challenge for our times. This does not mean indulging falsehood or corruption. It means moving from 'us/them' thinking to 'we' consciousness, from separation to cooperation, recognizing that we human beings are ultimately far more alike than we are different. This is challenging in situations with people whose views are radically opposed to ours. Be firm on the issues, soft on the people.

The practice of loving our adversaries is obviously challenging in situations with people whose views and methods are radically opposed to ours, but that is where the real growth occurs. As we discover that the problems of humanity are also found in our own hearts and lives, we realize that the 'them' we often speak of is also us. We are not exempt and we are not different.

As we discover that the problems of humanity are also found in our own hearts and lives, we realize that the 'them' we often speak of is also us.

Selfless Service of Others

Our work is for the world, not for ourselves alone. In doing service work, we are sowing seeds for the benefit of others. The full harvest of our work may not take place in our lifetimes, yet our efforts now are making possible a better life for future generations. Let your fulfilment come in gratitude for the privilege of being able to render this service, and from doing so with as much compassion, authenticity, fortitude, and forgiveness as you can muster. This is the traditional understanding of selfless service, and yet its opposite is also true, as reflected in the next principle:

Selfless Service Serves Us Also

In serving others, we serve our true selves. "It is in giving that we receive." We are sustained by those we serve, just as we are blessed when we forgive others. As Gandhi says, the practice of satyagraha ('clinging to truth') confers a 'matchless and universal power' upon those who practice it. Service work is enlightened self-interest, and it cultivates an expanded sense of identification that includes all others. So although we are not here to serve ourselves, nothing serves us better than serving others.

Do Not Insulate Yourself from the Pain of the World

Shielding ourselves from heartbreak prevents transformation. Let your heart break open, and learn to move in the world with a broken heart. As Gibran says, "Your pain is the medicine by which the physician within heals thyself." When we open ourselves to the pain of the world, we become the medicine that heals the world. If we push away the pain, we are actually preventing our own participation in the world's attempt to heal itself. This is what Gandhi understood so deeply in his principles of ahimsa and satyagraha. A broken heart is an open heart, through which love flows and genuine transformation begins.

What You Attend To, You Become

Your essence is pliable, and ultimately you become that which you most deeply focus your attention upon. You reap what you sow, so choose your actions carefully. If you constantly engage in battles, you become embattled yourself. If you constantly give love, you become love itself. Each one of us is entirely responsible for our particular life, and for what we choose to serve.

Take Sufficient Time for Retreat, Renewal and Deep Listening

Knowing when to retreat is part of knowing how to advance. The greatest spiritual leaders and activists have always taken significant retreat time away from the world, in order to better serve in the world. Sustained periods of conscious respite help to clarify vision and thought, relax the body and mind, and purify intentions. This expands our capacity to serve and cultivates a larger, more circumspect view of ourselves and our service.

Rely on Faith, and Let Go of Having to Figure It All Out

There are larger 'divine' forces at work that we can trust completely without knowing their precise workings or agendas. Faith means trusting the unknown, and offering ourselves as willing vehicles for the intrinsic wisdom and benevolence of the cosmos to do its work. "The first step to wisdom is silence. The second is listening." If you earnestly ask inwardly and listen for guidance, and then follow it carefully – you are working in accord with these larger forces, and you become the instrument for their music.

A foundation in unshakable trust is not Pollyannish fantasy or naïve idealism, as some 'realists' might interpret it. Rather it entails a deep and instinctive alignment with the mystery and wonder of life itself, invoking something real yet hidden that goes quite beyond traditional scientific principles. Faith is not blind adherence to any set of beliefs, but a knowing from intuition and experience about universal forces and energies beyond our direct observation. We can draw upon and engage these hidden forces, first by knowing they are there, and second by asking or yearning for them to support us – or more precisely, asking them to allow us to serve on their behalf. This realization actually brings great relief, as we recognize that it is not up to us to figure out all the steps to transform the world, because we are just participating agents in a much larger cosmic will and wisdom.

Love Creates the Form

Grace is ineffable to the senses, yet is no less real for being hidden. It is the power of love in action, and love is the greatest power in the universe.

Not the other way around. The heart crosses the abyss that the mind creates, and operates at depths unknown to the mind. Don't get trapped by 'pessimism concerning human nature that is not balanced by an optimism concerning divine nature, or you will overlook the cure of grace'. (Martin Luther King). Let your heart's love infuse your work, and you cannot fail, though your dreams may manifest in ways different from what you imagine.

What Martin Luther King calls the 'cure of grace' is fundamental, yet quite beyond what the logical mind can fathom. Grace is ineffable to the senses, yet is no less real for being hidden. It is the power of love in action, and love is the greatest power in the universe. On this point, King soundly refuted even the most compelling social and religious pessimists who relied on political and theological analysis alone (e.g. Reinhold Niebuhr's *Moral Man, Immoral Society*). To overlook the cure of grace is to overlook the very source and foundation of all life.

Epilogue

Before closing, it might be asked whether these principles could be 'proven' scientifically or philosophically in some way? Perhaps not in objective or empirical terms cherished by the materialist or the sceptic. However, as socially engaged spiritual activists, we do not live by our proofs; we prove by our living. In the words of the mystic poet Rumi, "If you are in love, that love is all the proof you need. If you are not in love, what good are all your proofs?"

As socially engaged spiritual activists at the dawn of the third millennium, we are called to serve in two distinct capacities: as hospice workers to a dying civilization, and as midwives to an emerging civilization. Both tasks are required simultaneously. The culture of modernity is disappearing, to be replaced with a civilization of love, if the human race is to thrive and prosper. We are called to move through the world with open hearts – being present to the grief and decay of a waning civilization – while at the same time maintaining heartfelt enthusiasm as we focus our energies on visionary inspiration and building unprecedented new forms of human community that will serve the future evolution of humanity.

Please see page vi for William Keepin's biography.

Findhorn ecovillage is recognised as a pioneering community that has successfully married spiritual, social, economic and ecological values and practices to create a community where people can develop fully. This interview with two Findhorn community leaders explores the richness of Findhorn's inner and outer life.

Silence and the Sacred:
Interviews with Craig Gibsone, former Findhorn Foundation Focaliser, and by Robin Alfred, Chair of Trustees

William Keepin

Drawing upon Findhorn's 44 years of experience, what are some of the key insights or lessons you would underscore for people today who seek to implement a spiritual vision in community?

The reason why I came to Findhorn, and the reason why I stayed, is that there's always been an acknowledgment here of the intuitive and the importance of silence – silence within oneself and the creation of a place of silence where you actually sit with your friends. From my Buddhist perspective and background, that is spiritual practice.

The importance of silence... "Be still, and find your unique relationship with the wider mystery of creation"...

What I try to get through to people today – particularly the younger generations who sometimes react negatively to the idea of meditation or spirituality – is that you need to practice silence, or practice being still, just as much as you need to practice being on your skateboard. Regardless of what your chosen interests and activities are in life, you need to put energy and time into being silent, and acknowledge its importance. That's central at Findhorn: "Be still, and know that 'I am... '". As one of our founders Eileen Caddy would say, "God is the still small voice within." Findhorn has also taught, "Be still, and find your unique relationship with the wider mystery of creation" and if you want to actually discover this, you must take the time do it.

Are there specific techniques for silent meditation at Findhorn?

There are lots of techniques for becoming silent. If you want to be a good mountain climber, you need to put yourself out there in the steep terrain and practice climbing but if you want to practice sitting still and listening, then you need to work with your breath. It's the best way. Simple. In that context, we teach very simple rhythms of breathing, like a rhythm of dance.

I've been impressed with Findhorn in that it hasn't developed any real dogma or specific techniques; the basic practice is just to 'be still'. Each person discovers their own unique path to becoming still. And it's important that a spectrum of methods exists to choose from, because different people require different pathways to discover their inner silence. While I love sitting meditation, I realize that some others actually love jogging, or the mountain bike. But regardless of the method, it's important that your method takes you to the place where so called 'Zen moments' happen. Whenever I teach, I try to ask 'what is the best tool for each one of you?' I encourage each person to inquire, "What is my individual path?" Findhorn's longevity is, I believe, that it has been incredibly accommodating in holding people and supporting them to find their own particular rhythm and space. At Findhorn, a very deep quiet intention has been established – not dogmatic or clearly written down. When you walk into the sanctuary, for example, you sense and feel that you need to sit down and be silent there.

A key to Findhorn's success is that it has been willing to embrace some very powerful techniques for inner work, including methods for entering so-called altered states or transpersonal states of consciousness, or states of visionary receptivity or intuition. Although we draw upon these methods, Findhorn has never become a vehicle for their sole expression. For example the Arcane School (of Alice Bailey and Dwajl Khul) has played an important role here, but it has never predominated. Yet there's always been a group that practices that particular form, holding a strong inner focus and discipline. Other groups in the community are committed to other paths, such as Buddhism which has been another building block here.

This spiritual diversity at Findhorn has been one of the strengths of the community and it's very purposeful. Findhorn shows a way into the future that is inclusive, eclectic, open minded and very tolerant. For example, it's in the practice of attunement, which happens every morning in every work department. People come together and spend a few moments just being together. And, most important, they share what they're feeling, and what their particular needs are.

The practice of 'attunement' came in strongly in the 1970s, and it enables people to be present with each other and share what's happening on several different levels. For example, it might be someone saying, "I just had a major confrontation with so-and-so this morning, and I'm deeply disturbed." And this comes up before we ever start cooking the meal, and that person's distress is 'held' sensitively within the fabric of the work department that day. So at Findhorn, very human needs are being met all the time. And then at times, all that is set aside, and there is just open space. We are rather Quaker like,

Findhorn shows a way into the future that is inclusive, eclectic, open minded and very tolerant.

and in fact Quakers are often attracted here, because they too acknowledge the importance of shared silence.

Findhorn also has various elements of Buddhism. For example, work is love in action, which is mindfulness. Then there is a strong emphasis on service: giving and sharing without thinking of monetary reward, or any level of reward, even just being congratulated.

What is the locus of spiritual authority at Findhorn? Did Eileen or Peter serve as 'gurus' or spiritual leaders? How has spiritual leadership evolved here over the years?

A primary characteristic of Findhorn's spirituality has been the recognition that you don't need some level of officiate, or guru, or someone who mediates between you and That. But you do need the humility and compassion to recognize that we are not alone, and that there is intelligence imbued in everything – always present, always changing. Another key aspect at Findhorn has been the concept of co-creation with nature, which developed particularly in the 1970s. Human beings are nature. You are nature; there is no separation between you and nature. Even the sound of that jet screaming across the sky through our daily lives all the time! [Findhorn is located adjacent to an air force base]. I've come to the point where the jet is the sound of nature, a product of life on earth evolving, and we are that. Just as other species make their particular noises, we humans make our particular noises.

At Findhorn, we are always open to the unseen – always acknowledging that the seen and the unseen are ever present. There are simple little exercises that we still do, in which you take someone's hand and ask them to close their eyes, and you escort them through the garden, and they are given something to touch with their eyes closed, and they learn to 'listen' with the senses. I think that through our shamanic work and other transpersonal states, we are sensing and feeling that 'nature' is inside us as much as it is around us and through us. So the unseen is always present.

Findhorn is famous for working closely with nature spirits, for example, growing 30 pound cabbages in the challenging climate of northern Scotland. How has this practice unfolded over time at Findhorn?

Co-creation with nature is one of our strongest attributes at Findhorn, particularly now as environmental concerns have become so important. To become sustainable, all of our 'harder' technologies have to become organic and integrated into the natural environment. We don't have the 30 pound cabbage of the 1960s now, but we have a 'living machine' that processes our sewage and produces wonderful plants and clean water at the back end. People may object that it's just a collection of tanks and pipes and plants, and we say yes, but all those life forms in there we are co-creating with one another. When it's blooming, you could say it's living off our shit, but those teeming lifeforms in there are actually celebrating, saying, "Oh yes, how wonderful, fantastic, beautiful. Please more!"

So we take people through the living machine and show them how it's mimicking the story of creation. As you enter the first door, it's like the first life on earth. There you have the anaerobic life and sludge. Continuing through the next door, all of a sudden oxygen starts to be produced, and you have the second great wave. You step through this door of the living machine, and you're in the first great wave of extinction. As you continue onward, it is mimicking the whole of natural evolution in there, albeit without the dinosaurs. As you keep going, you come to the sixth great wave of extinction, and you stop right there and ask the question: Who is the species that is involved in this great wave that is beyond any of the great waves of past extinctions? And it's us! Humanity! And so the living machine touches what I call my biological soul. I don't know if many people understand this, but it's my primordial nature; it's that little organism that was me all along, and which has slowly evolved and grown until it has become this extraordinary powerful living being we call the human being. And we humans may in fact choose that path.

From the living machine we next go to the Universal Hall, down in the basement; or to the sanctuary in a human constructed environment; you dim the lights, and you're in infinity. You're in a temple. And that is my soul. So the one is my 'biological soul' and the other is my soul. And maybe that's what people feel when they come to Findhorn, because nature does give sustenance. They may not recognize it consciously, but they also feel nurtured because there is this atmosphere of caring and love here. There is willingness for silence but also willingness for relation and intimacy.

Another important dynamic of this place is reflected in a conference held here that was called 'Psyche and Soul'. The psychological aspect of ourselves is also acknowledged as sacred. The Dalai Lama expressed it perhaps best: when we go deeply into spiritual search, and particularly for Westerners who go into meditation and silence, we need psychodynamic skills to process the material that is coming up and arising through us. Findhorn does this through various means – Holotropic breathwork, co-counseling, the Game of Transformation – all these are tools necessary for continuing one's personal composting.

What are you most passionate about in relation to Findhorn now?

I'm fascinated by the observation that, starting out from a spiritual perspective, the deeper I went into the spiritual nature of myself and the mystery of life, the more I was led to nature and the environment. The mystery must be there, as well as inside of me. I am part of nature. What is it really like? Can I sense and know it directly? How does an aboriginal actually perceive time and space? If you ask aboriginal children to draw self portraits, they will draw the environment, the trees and the landscape.

What I'm passionate about is how the environmental movement is coming closer and closer to spirituality. In our ecovillage trainings at Findhorn, for example, we're seeing people coming from a spiritual background and an environmental background, and they're crossing over into one another's territories and shaking hands. In the ecovillage movement, these people are working more closely and intimately with one another, infusing one another with their respective languages, wisdom, and sense of meaning. They can both give one another a deeper sense of what true listening is. The power of the ecovillage movement is that it's creating a rich ground for cross-fertilization, for people to share from diverse perspectives and constantly observe. This process is very strong in the ecovillage training here. Times of silence are written into the schedule. No task is so important that we haven't got time for silence. Whatever the educational processes are, they must be built around the life and the nurture, care, and flow of the community we are operating within. Otherwise we get nowhere.

How are children nurtured and educated at Findhorn?

I have never seen this community attempt to raise our children by imposing a particular way upon them, e.g. "The children need to meditate, or they need to communicate with nature," or any of these things that we all do. We just allow them to experience it, to see us work out our differences, and also to see us celebrating, singing and dancing – which at points they really love and other times they find very embarrassing! But for children and youth here, Findhorn is always a compassionate loving space that allows them to move through all their experiences. They really love it and enjoy it in the early stages, and later when they come of age they often reject and leave it and go out to discover the larger world. Then often they come back, and regard this way of living a very good thing.

How is the inner strength of a community sustained over time?

The key factor in any attempt to create a community is that you must feed and nurture the vibration that you want. You don't have to call it spiritual. What is the presence that you want there? You have to constantly work on that; invoke it, invite it, and consciously ask for its participation. You don't have to do this in 'spiritual' terms or language, you can do it shamanically, or even in other very simple ways. It's like a 'group discovery' process that brings you into a place of listening, and it doesn't have to be completely silent. But you need some way for people to show the ultimate sign of respect for one another, and that is to sit together in silence... eyes open or closed, whatever is comfortable.

When someone becomes obnoxious in their ego attachments, or gets 'out of line' in their behavior, how does the process of self-correction function at Findhorn?

The founders were really good in setting up a self-regulatory type of process that helped us to monitor one another (and among the 'founders' I include here less visible ones like David Spangler and others). In the early days, we had TLC (tender loving care). Whenever we saw someone in a disturbed space, or starting to act 'differently' in an inappropriate sense, we would pay special attention to him or her. This has evolved over time into what we call S&PD (spiritual and personal development), and it entails a learning role. One thing Peter was really good at was shifting people through different jobs and responsibilities, so that people were not always doing just one job. Over time, people would experience a whole range of different roles, including simple jobs like just cooking in the kitchen. But this too might grow over time into focalizing the kitchen, which means cooking for 150 people.

This created a rich process of self-monitoring that would and guide and help a person to grow. The self-monitoring also requires working with one's emotional and psychological bodies; learning what they are, how they manifest, and how to identify one's own experiences. As people gain some level of self awareness and insight, they can start to support others in this way as they navigate through different areas of responsibility in the community.

Spiritual monitoring is of course part of it too. There is a lot of power and charisma projected onto people who can work with so-called spiritual issues or with difficult people. At Findhorn the great thing is that we're all expected to work with these issues. We don't all have to end up specializing in this arena, but we all work with spiritual emergence, and with spiritual emergency. Every now and then someone has a crisis, and there is a need for more intensive care.

When you enter into the Foundation, you are mentored or monitored, and the basic process is really straightforward: you set your own parameters or goals for where you want to get, and the monitor's job is to check in with you periodically, and see out how you're coming along in relation to your own goals. Then whenever you are in transition of some kind, help is available. So for example, when someone feels complete in some area and is ready to move on to something else, the support is there. The 'classroom' of the community is very fluid.

Whenever someone gravitates toward one of the more glamorous positions, and perhaps become a bit ego-inflated about it or starts telling everyone what to do, we immediately bring that person back to reality. Where does this humility come from? I would lay it back at Eileen's lap, when she received guidance that she would no longer get guidance for the community. This was a key step, because then the community had to go inward for its own guidance, out of necessity.

As people gain some level of self awareness and insight, they can start to support others in this way as they navigate through different areas of responsibility in the community.

How is the Findhorn community deepening and sustaining its roots over time?

What I see is an amazing amount of dedication and commitment to the founding principles of the Foundation, it's almost like a lineage. There was a vibratory field set and held by Peter and Eileen, and people came in through that and in through them. This process slowly grew to the point where there is an extremely strong presence now, particularly in the Findhorn Foundation as an entity, and you can sense and feel this very strong body of light. And people are initiated and welcomed into that by those who already have been, and also at times they exit this field. So there's an invitation to move into an energetic field that is held by the people who have already been initiated into it.

This deepening goes on and is always present, even as the outward expression of it changes. People often say that a lot of change has taken place at Findhorn over the years, and yes it's true – physically – but from a vibratory level, it's exactly the same space. I trust my buddies who are always holding that space, who actually chose to say yes to this community, and hold this space together. And each one of those people, to me, are as powerfully and deeply evolved and connected as Peter, Eileen, and Dorothy were.

Formerly, there was more power and charisma embodied in one individual, rather than in many. I think it's still important to have the Eckhart Tolle's and other wonderful inspirational individuals who can be like the Dorothys and Peters for us. But at the same time, I see people in the community doing it for themselves now. I've traveled with Peter, Eileen, and Dorothy and seen the way they teach and work and deliver around the world, and now I travel with other people and see them doing exactly the same thing, and I myself go out and do the same thing. Except now it's not just one or two people, it's dozens. There are 40 or 50 people at Findhorn who are constantly traveling, offering a broad range of community skills and knowledge such as sacred dance, environmental education, musical workshops, leadership and conflict resolution trainings. They give the same level of work at the same level of quality as Peter and Eileen and Dorothy did in the early days.

Earlier you mentioned the Arcane School (teachings of Dwal Kuhl presented in the writings of Alice Bailey). What is the role of this or other esoteric traditions at Findhorn?

The level of intelligence and wisdom embodied by the Arcane School is very powerful, and there are several people in the community who plug into that channel and bring it into the community. Some are deeply and almost religiously dedicated to a particular dissemination of energy from esoteric sources into the wider body of humanity. This was perhaps part of Eileen's final work.

Could you speak of relations between the masculine and feminine at Findhorn?

Findhorn is definitely a place where the feminine and masculine are reaching a place of equality that I believe is unprecedented. Gender balance is a key issue for any emerging ecovillage: a genuine acknowledgement and sense of total equality is crucial, in which masculine and feminine are viewed as different yet equally vital and necessary. This is where the emerging ecovillage differs greatly from a traditional village. The latter often has many key elements that we also strive for in terms of sustainability, but the gender dynamics in traditional villages are usually far from balanced.

And I know some feminists would say, 'Ah yes, but, Findhorn has a long way to go.' It's true that we're not there yet, we're not liberated to any great degree. Yet Findhorn is based on the intuitive and receptive, and that's why the power of the Feminine is strong here. Just about every major position in the Findhorn Foundation is held by a woman at this point. There are one or two men in these roles, but all the other leadership roles are currently held by women. At Findhorn the feminine is always consciously invited. We want that part of you, as much as we want that part of you that is the man, the 'practical' part. But what we really love to see is a man comfortable with being feminine within himself, however that may manifest. So in any ecovillage, a key aspect is to bring the sexes into a place of co-creativity, just like co-creativity with nature. We need to have this integration within our sexual identities.

So in any ecovillage, a key aspect is to bring the sexes into a place of co-creativity, just like co-creativity with nature.

Are there many gay, lesbian, or transgender members of Findhorn?

Very many. The gay feminine is especially strong and present here. What I'm also seeing with the new generations coming through now is that they don't identify with a particular sexual orientation. They are sexual, that's it, and they want to stop this business of putting their sexuality into a box (such as hetero, gay, lesbian, transgender, or bisexual).

Is there anything else you wish to say by way of closure, or concluding insights?

Yes. Wherever a new initiative is begun, whatever it's called, it is vital that the initiators or project leaders are aware of the soul of the land and the traditions of the land which must be allowed to somehow inform and infuse the project's evolution. From a Findhorn perspective; it's not an accident that our community was born in a Celtic tradition and land, even though much of that heritage is lost. It was very much a land-based, nature spirituality. Ecovillages are here to restore and nurture the balance between the four elements – earth, air, fire, water – and between nature and humans. This is a key aspect, and people become aware that the land and the things that have passed have always been speaking through the people. This leads me to Chief Seattle, "Everything is sacred unto my people." And in that, it's not just the natural environment that is sacred. My emotions are sacred unto my people. Everything that is taking place is seen as an aspect of the sacred. Seeing this is the essence of Findhorn.

Robin Alfred, Chair of the Findhorn Foundation Trustees adds to the above in conversation with William

Could you briefly outline the spiritual philosophy of Findhorn, as it was in the beginning, and as it is now?

Findhorn was founded 44 years ago, and the three founders – Eileen Caddy, Peter Caddy, and Dorothy Maclean – all came with strong spiritual lineages and backgrounds. Eileen was a meditator, part of the 'moral rearmament' movement. Her focus was on inner listening – "Be still and know that I am God" – listen to the God within for answers and ask for guidance within. She spent a lot of time in the sanctuaries, creating space for herself to meditate and receive guidance, listening for the 'still, small voice' within, which is her practice. Peter was a man of intuition, with a background as a Rosicrucian, who also served in the Royal Air Force. He was a mountaineer and climber; very much an intuitive who brought things into action. His principle was the principle of will and action, but based on intuitive knowing, and informed by Eileen's guidance. Dorothy was a Sufi when she came here, and her practice was listening to nature, and cooperating with nature's intelligence. These three practices – silent meditation and listening within for guidance, tuning into and cooperating with the elementals and the devas, and bringing these things out into action on the human level – are still the cornerstone of spirituality at Findhorn.

Because Peter left in the early 70s and Dorothy also left in the 70s, the strongest practice today at Findhorn is silent meditation and inner listening. This is because Eileen was the only founder who stayed here. She died aged 88 in autumn of 2007. The other two principles, the more masculine principle of the will, and co-creation with nature are weaker in the community. Dorothy has been doing a lot of teaching recently and has been visiting the community to help us explore how strongly we are focused on co-creation with nature today. Peter's leadership principle was more hierarchal and is probably more difficult to manifest in contemporary Findhorn, because people conflate the 'inner' principle of will and action with the outer 'male' experience of leadership and hierarchy, so there is some resistance to that I believe. Of the three founding principles, it's the weakest principle that's alive here now.

What are the contemporary forms of spiritual practice at Findhorn?

There is a huge spectrum. We have a spiritual practice directory that lists some 45 different spiritual groups that meet here, spanning a wide range of practices. Buddhism is strong, including both Tibetan forms and Vipassana practice. Then there is yoga, two different Sufi practices, a Course in Miracles, as well as more psychological practices like Non-Violent Communication, and Five Rhythms as a dance form, and many other modalities. Beyond these more specific forms, there are several core practices that are open to all, and practiced every day in the community. These core practices would include: meditation at 6:30 am for an hour, collective meditation form 8:35 to 8:55, Taizé singing from 8 to 8:30 every morning, blessing of food before each meal, and attunement. Not everyone does every practice, of course; it is left up to individual choice.

The attunement practice is key at the beginning of events and conferences, as well daily in the workgroups, and it's used to mark the beginnings and endings of virtually everything here. I work with a leadership training process called mytho-drama that draws upon the wisdom in Shakespeare's plays as a template for creating leadership training seminars. We run these seminars for corporate executives, and we often hear afterwards from these busy executives how when they went back to their normal working life and offices, how much they missed the soulful attunement practices of Findhorn. Or in another context: in our introductory Experience Week program I sometimes facilitate a sharing process on personal and planetary transformation, and even just these simple things make a huge difference. In most places it is unheard of to start a meeting with a check-in – to sincerely explore how people are doing personally, emotionally, and, with permission, spiritually. Yet this puts people into relationship before they rush off into their business, and it makes a huge difference. It brings tangible connection, or heart connection, and warmth starts to circulate in the work place. This is a core practice at Findhorn.

In most places it is unheard of to start a meeting with a check-in... Yet this puts people into relationship... It brings tangible connection, or heart connection, and warmth starts to circulate in the work place.

Given Findhorn's more than four decades of community experience, what are some of the major lessons and insights you would emphasize for those seeking to implement a spiritual vision in community?

You can't do it alone. The experience of building groups and meditating collectively, as well as on one's own, is very helpful. Peter Hawkins recently came here as a spiritually awake consultant, and he identifies four kinds of meditative practice: (1) meditatation on your own for yourself, (2) meditatation on your own for the collective, (3) meditatation collectively for yourselves, (4) meditation collectively for the collective. And it's very useful to distinguish these, because for example, if you sit in a group meditating for yourself, or you sit in that same group meditating for the collective, it's a qualitatively very different experience.

So there needs to be space given to one's personal practice, whatever that might be, such as spending time in nature or accessing intuition or meditating, AND there must be time given for spiritual practice on behalf of the collective (whether done on one's own, or in the collective). Both these

Living in community, and creating space for honest feedback and reflection of what we see in each other is another key.

dimensions are needed, and I think that is a major learning from Findhorn. We not only develop ourselves spiritually, we do spiritual work on behalf of the collective. The group soul or group being is nurtured through some kind of collective practice.

Living in community, and creating space for honest feedback and reflection of what we see in each other is another key. In the absence of a spiritual authority figure, who will 'call you on your stuff' when it arises, we have to call each other on our stuff. Because it's very difficult to distinguish the 'still small voice within' that is God from the 'still small voice within' that is your most persistent, deep seated desire! We can do our best to distinguish these things but it's also very helpful to get honest feedback and reflection from people around you, and colleagues in the community. Setting up systems where we can be transparent about our process, and share what we're experiencing and get substantive feedback is really helpful.

Has Findhorn done this well?

Yes, I think we have done that quite well. Every work department has an attunement once a week for half a day, where we meditate collectively, share business, etc.. In its high form, it's an opportunity to share about one's personal journey and get other people's feedback about what they see in you. In its lower form, it can just become a personal sharing, and people respond by saying 'thank you for sharing', and there's not much challenge or feedback. Because feedback is key: honest, robust, loving, tough, compassionate feedback. It's very helpful for spiritual growth, particularly in a place like Findhorn where it does not come from a recognized spiritual master or guru. We've recently also imported the 'Forum' from ZEGG, which is a strong process for transparency and feedback in community.

In 1973, Eileen received guidance that she was not to share her guidance anymore with the community (including Peter), in order that the people in the community could get their own guidance. She continued to get guidance, but was told not to share it any more. It was a radical change in the community. Before that, people would gather in the sanctuary and Eileen would share her guidance for the day. Some of it was very broad, some very specific, and she often gathered this guidance by staying up for hours in the middle of the night. Whatever it was, Peter would translate the guidance into the actions for the day, and Dorothy would integrate the nature spirits and intelligence. When Eileen stopped getting guidance for the community, the community had to grow into the ability to tune inwardly and get guidance for itself.

Prior to that time, was Eileen in the role of a spiritual guru who would give people feedback?

No, Peter was more in the role of the one who would, for example, yank people out of bed and say, "You have to go to sanctuary – it's a requirement here." He would drum people out of the community if he felt they weren't the right person, and direct people to different tasks and departments. He was very much the challenger, and he did a magnificent job of running the community. I think when Peter left, a degree of challenge also left with him. And some people resisted Peter, of course. They felt it was time for the community to grow into more of a democracy with consensus decision making.

What is the current governing structure of Findhorn?

It's basically a series of concentric circles, if you will. At the hub are the Findhorn Trustees, and beyond that is a Management Team of about 12 people who are selected by the coworkers. Beyond that is a Council, consisting of about 40 people who are self-selecting, and they meet regularly, stay informed, read all the relevant documents, and then take decisions in the long-term interests of the community. Beyond them there's a larger body of co-workers who are not part of the Council, and beyond that there is the wider Findhorn community which is not part of the Findhorn Foundation, but is connected to the same impulse. It's important to distinguish between the Findhorn Foundation and the Findhorn community in its entirety. The former includes about 200 members, and the larger satellite community is about 500 people.

What is the relationship between Findhorn's work and the esoteric traditions, as reflected, say, in Alice Bailey's writings, or the ancient Masters of Wisdom?

I wasn't here in the early days, but my sense is that this element was quite alive at that time. Alice Bailey's writings, St. Germain and the Western mystery traditions were taught consciously in the beginning. It's not as strong now. In the 10 years I've been here, I've never been to a single teaching about that, although I do hear about it. The teachings I received here were much more about how to meditate, how to listen within. Eileen had very little time for the esoteric material, frankly. Peter would give her all sorts of books to read, but she told me that she never read all that stuff.

In the absence of a recognized spiritual leader or guru, where is the seat of spiritual authority at Findhorn? And how does the Findhorn community address its 'shadow', i.e. those inevitable dynamics and issues that become somehow denied, hidden, or suppressed?

At Findhorn we embrace the shadow, and work on its transformation by applying some of the more psychospiritual practices, such as Process Oriented Psychology (Arnie Mindell), psychosynthesis (Roberto Assiogiolli), the Forum (ZEGG), the Game of Transformation, and others. Sometimes we do this kind of work on a community-wide level.

The question of spiritual authority is a deep one. As Trustees, we wrote a resolution a couple years ago that asked the Spiritual and Personal Development branch of the Findhorn Foundation to conduct a six month inquiry into 'spiritual authority' in the community. The response from most Findhorn community members was that they are being trained to receive their own guidance, and they are reluctant to place spiritual authority in an external figure. This widespread response is partly due to the particular training here which emphasizes each person's need to get their own guidance. And this is why feedback is so important, as a way to guard against the potential pitfalls of relying solely on inner guidance. As for spiritual masters, they are not entirely absent. Through the 45 different spiritual practices at Findhorn I mentioned earlier, we have access to true spiritual masters. For example, Marshall Rosenberg was here recently teaching NonViolent Communication and people who went told me they were in the presence of a spiritual master. The same goes for Eckhart Tolle who comes and teaches here, or Carolyn Myss and others. So recognized masters do come through here periodically, and our training and practice is to develop that mastery in ourselves.

Yet questions of spiritual authority and accountability need to be kept alive always. There are lots of easy answers, like "I'm accountable to God." But what does that mean in practice? Who will call you to task when you're actually being accountable to your own ego? Because the ego is an extremely slippery thing, and it will quite happily pose inwardly as your spiritual authority and guide! So this question is inherently challenging, and we keep it alive at Findhorn, all the more so because we don't have a recognized spiritual guru in place here.

Another important perspective on this question comes from the 'incarnational spirituality' of David Spangler, who was one of the pioneering leaders early on in the Findhorn community. David has also reconnected with the community lately, and I recently did a workshop with him. At one point he drew two large circles on the floor, one representing the 'circle of our souls', and the other representing the 'circle of our humanity'. He said the fallacy of spiritual life is that people think what they need to do is to live in this soul circle. But in fact, what we need is to live in the fusion of our souls with our humanity. In the humanity circle, we are given all our mundane 'stuff', our social conditioning, feelings and emotions, disease and aging because we are incarnate beings. And this is as much a part of who we are as the soul or spiritual essence. I believe David would say that the proverbial battle of the 'higher self versus the lower self' is wrongly framed, because it implies that we have to get rid of these mundane human things, which is not the nature of our task. Our task is not to leave the humanity circle to live in the soul circle, rather we need to meld the two together, fusing soul into humanity. And that's how the Findhorn community operates.

There are some 45 people here engaged in on-line study courses with David Spangler. David says that we have a personality, and rather than it being something we need to transcend, it's a God-given reality that we have a personality. I asked David a question about spiritual teachers:

Our task is not to leave the humanity circle to live in the soul circle, rather we need to meld the two together, fusing soul into humanity.

"In the absence of a spiritual teacher, how do you dismantle your own ego?" He replied that this is old thinking. The notion that there is a 'sun-centered' universe, with disciples orbiting like satellites around the guru, is old thinking. The new way – of which Findhorn is seen as the exemplary grandmother of new-age communities – is that everyone is in relationship to everybody else. Everyone is learning and giving and receiving, and it's not a sun-centered universe anymore. We are all suns. This is much more the Findhorn way of thinking. It's not about gurus, it's about the God within, collective consciousness, calling each other to our highest, and holding each other accountable for our 'stuff.' We are now entering the Age of the group, rather than the Age of the master. The new 'Buddha' will be a group rather than an individual. This is the Findhorn paradigm, and we attract people who resonate with that.

But surely these two perspectives are not mutually exclusive? For example, the Tibetan Djwhal Kuhl says just what you're saying about groups and communities, but at the same time he fully upholds the pivotal role of illumined Masters. And indeed, if David Spangler has 45 people at Findhorn studying with him plus many more in the United States, doesn't that make David a de facto sun, around whom his students are orbiting? So just because this is the Age of the Group does not mean the Master vanishes, does it?

No, I don't think so. Perhaps the role of the Master changes. But these are deep questions, and David Spangler, along with William Bloom and others, is the arch-exponent of the way-without-gurus that Findhorn upholds. At Findhorn, we don't talk about the annihilation of the ego, or the dissolution of the ego. We explore how to bring your humanness and your personality into spiritual relationship. We look at how you wed your soul and your personality.

At the same time, it's hard to deny that there do exist at least a few highly advanced spiritual masters whose depth of realization and awakening far exceeds that of the majority of spiritual seekers. Indeed, perhaps David Spangler is one of them!

Yes, and perhaps the new thinking complements, rather than replaces, the old thinking.

Is there anything else you would like to add by way of closure or conclusion?

Yes. At the risk of oversimplifying, community is the answer. Community is the way forward, and I believe that the emerging ecovillage paradigm is one of the most important developments that can help save the planet.

Thank you, Robin. This has been a rich and very illuminating conversation.

Craig Gibsone was born in an isolated farm in Australia. He is a builder of practical organic structures, specialised in retrofitting houses to passive solar. He has lived for the last two decades at the Findhorn Ecovillage where he was involved in the construction of many community buildings, including the Community Center and the Barrel Houses. He is an artist and musician and teaches in the International Holistic University. Craig works internationally as ecovillage building consultant.

Robin Alfred worked as a trainer, educator and social work manager for 15 years in London, prior to coming to the Findhorn Foundation, Scotland in 1995 where he has served as Chair of Management and Chair of Trustees. Robin is a faculty member of the Foundation's Ecovillage Training and has taught on EDE courses in the UK, Germany and India. He is co-editor of *Beyond You and Me*, the Social Key in the EDE curriculum.
www.findhornconsultancy.com

Nestled in a valley in the Piedmont hills is Damanhur, an 'ecosociety' of 1,000 residents. Macao Tamerice explains how this ground-breaking society is organised to express its spiritual values.

Spirituality in Damanhur

Macaco Tamerice

Founded in 1975, Damanhur is an ecosociety comprised of about 1,000 citizens: a federation of communities and ecovillages with their own social and political structure in continual evolution. Damanhur is a center for spiritual, artistic and social research. It's philosophy is based on action, optimism and the idea that every human being lives to leave something of themselves for future generations and to contribute to the growth and evolution of the whole of humanity.

In Damanhur, respect for life in all its forms is a basic tenet of its philosophy. Damanhur was created as a place where it is possible to live spirituality 24 hours a day – to bring matter and spirituality into harmony. Damanhur's concept of spirituality is very broad: give meaning to things and use positive thought to direct the best of our energies. From this point of view, Damanhurians consider everything one does as spiritual: whether they meditate or pray, conduct research or art, when at work giving themselves to the others, when taking care of the children or simply cooking or cleaning the house.

This respect for life fosters the wish to share deeply with other human beings and create community. By finding new ways of living together and using others as mirrors, individuals are encouraged to develop inherent talents; each person learns to understand and express who he is through a deep interconnectedness with others.

Understanding that diversity is an expression of the different facets of life that we all represent helps us to see differences as enriching rather than an element of separation. In this way, living together is a daily exercise of spirituality because it means raising the quality of the relationships with others in order to better ourselves and the reality around us.

Living together is not always easy; sometimes our strong ideas clash with the strong ideas of others. The first reaction is often defensive, making acceptance of diversity a difficult tenet to live up to. In Damanhur there is the concept of 'quasi reality', which starts with the assumption that there is no objective reality. Everybody has a personal interpretation of reality, so our perception of what is real is always colored by our own personal experiences and convictions. Why does this matter in moments of conflict? It means that on the one hand, of course "I'm right", but I also know that the other person is right too. It is not so important who is right or wrong, what really matters is to find a solution that works for the two parts.

Community Organization

Today, of the roughly 1,000 citizens of Damanhur, about 600 are full time residents living in nucleo communities of 10 to 25 people. The rest are non-resident citizens who live in their own homes and participate at various levels of their own choosing.

The way of organizing the communal living of the residents has changed many times over the years, like everything in Damanhur. One of the few Damanhurian constants is that everything changes.

When Damanhur was first coming together, the organizational structure of the community had to support quick decision making. The founder and inspirator of Damanhur, Falco – Oberto Airaudi – became the governor and five Damanhurian ministers were appointed. This system was in place for less than two years, after which the community found itself strong enough to move from a centralized organization to the democratic system seen today.

The King/Queen Guides, usually two or three people, are elected by all Damanhurians every six months and can be re-elected indefinitely. As the name suggests, they are the people responsible for Damanhur as a whole and are called King/Queen Guides to express the spiritual nobility of their role.

Falco no longer holds any official role inside of the Federation. Today, he is focused research in 'spiritual physics' and shares the results of his work through weekly, public question and answer sessions with Damanhurians and guests. One of Falco's strongest qualities is the ability to make people dream. His teachings aim to help people find the master within. Through the years he put forth many proposals that proved to be very successful. For that reason his ideas are considered highly when he brings proposals and comments to the community.

Damanhur is a Federation of communities where every community takes care of a specific task inside of the Federation. (The designations community, nucleo and family are synonymous.) There are groups that research agriculture, others in the fields of education, renewable energies, woodland management, and so on.

Obviously as far as all general fields are concerned, they are present as well in all or almost all the other groups, but the group that has the task spearheads the research in their specific field. The results are shared throughout the community, which allows the Federation to research many subjects simultaneously.

Individuals choose the family group that they want to live in according to its projects and synergies with group residents. As every community of the Federation has a specific task for the whole, every individual living inside the communities has a specific role, just like in a big family.

One person is elected as representative of the family every year, based on a program they propose during the election, whilst another person takes care of the finances, and one does the shopping, etc. The groups meet weekly to discuss projects, the organization of the house, personal achievements and challenges, and what is happening inside of Damanhur.

Two or three group form a region and elect, once a year, based on a program, a representative called 'captain'. All the captains meet weekly with the King/Queen Guides.

During the assembly between the captains and the King/Queen Guides, the captains bring ideas and proposals from their groups, whilst the King/Queen Guides share information about Damanhurian achievements, challenges, news, and ask for opinions and elaborations from the families. This allows information to flow constantly from the bottom up and the top down. It gives the families the opportunity to be heard and provides the King/Queen Guides with a continuous updates about what is happening in the community. This system enables the Guides receive continuous feedback and ideas from Damanhurians, whilst the citizens are kept informed about what is going on across all of Damanhur.

In Damanhur, the more power you have, the more you are in service. For every office held, at the end of each term, there is a public meeting to review the person's performance based on the program proposed during the election. In this meeting, the person who holds the role updates the status, celebrates achievements and evaluates failures, exploring next steps, what was learned as an organization, as a person, and so on. Members of the community ask questions, make comments and express their opinions.

Creative Living

Another really important theme in Damanhur directly linked to the idea of spirituality is art. When you create any form of art, you are in touch with the creative parts inside of yourself. It means to take inspiration from our best parts and transform reality around us. This idea has deeply inspired the community since the beginning. The belief that every person is an artist has given many individuals the encouragement to develop their artistic skills and talents, creating something beautiful and meaningful for the others.

Art is expressed at individual and collective levels. On the individual level, it means to explore ourselves and give expression to our values and talents... to dare to use our imagination and go beyond our perceived limits and habits.

Damanhur's idea of art is not linked to the expression of suffering or

The belief that every person is an artist has given many individuals the encouragement to develop their artistic skills and talents, creating something beautiful and meaningful for the others.

to conflict, but to our opening to life and to accept and give meaning to what life puts before us. This type of research opens a channel to the divine, whether the divine within everyone, or a transcendental idea of the divine.

Damanhurian philosophy emphasizes that there is a divine spark inside everyone and everything and that there is a divine force comprising all of existence, each part being a manifestation of an all-pervading force called 'God' in all philosophies. Between this divine spark and the all-encompassing divine force, according to Damanhurian philosophy, there is a whole ecosystem of divine forces creating many different cultural expressions of the divine. Within Damanhur's concept of spirituality, there is space for all beliefs as long as they are expressed harmoniously. They are regarded as cultural manifestations of the divine in different moments of time, and the essence of the many different peoples that have existed and exist on earth. The spiritual path of Damanhur, called 'Meditation', seeks contact and reawakening of this divine spark within.

The Temples of Humankind

On a collective level, art means expressing the values of a group of individuals, a people, towards something bigger. Humans groups seem to feel the need to create a sacred space in which to meditate or pray, to be in touch with the sense of divine lying inside every human being. This is also true of Damanhurians. A huge underground building, an artwork, dedicated to the reawakening of the divine essence in every human being was excavated in the mountain and called the 'Temples of Humankind'. It is considered by many as the 'Eighth Wonder of the World' and will celebrate its 35th anniversary in 2013.

The Temples of Humankind have become the expression of the creative power of the Damanhurians. One of the canons of Damanhurian art is to work together in a group. To understand this unusual idea of creating art we can look at the building of a hall of the Temples. Every time before building a new hall the builders, technicians and artists met to first explore together which values they wanted to express in this hall. After giving birth to the general idea, the different art laboratories took over a specific part. So the values were already expressed during the building process through architecture and technical refinements and then were realized more specifically through the different art laboratories. For example the painters made a sketch and then painted the hall together; or the painters co-operated with the mosaic makers created a sketch for a pavement in mosaic, which the mosaic makers then made.

The principal idea of Damanhur's concept of art is therefore on one hand the expression of ourselves, on the other the idea that we can transform reality by creating something meaningful and new together within different groups. Every Damanhurian is both artist and audience, infusing one's work with a sense of timelessness and spiritual significance.

Every Damanhurian is both artist and audience, infusing one's work with a sense of timelessness and spiritual significance.

Nature and the Spiritual Worldview

The spiritual approach to life at Damanhur affects many areas, especially in the interaction with the environment: if a deep respect of life pervades your worldview, how can you not be ecological? Since the beginning, there has been a push toward organic products and sustainable living. For over 35 years, waste products have been recycled, even at a time when there was no recycling in Italy, because it was important to Damanhur to create awareness around waste.

Another way to express our closeness to nature is through our names. Every Damanhurian chooses an animal or mythological creature for a first name, and a plant name for the second name. Changing names is a big life change, a change of identity, and this change mirrors the deep transformation that happens on the Damanhur spiritual path.

Changing your name changes your frequency. Living in community shows the importance of being available to change and transformation, and a willingness to discover new parts of ourselves in order to become a better person. In this sense, changing your name encourages transformation. Of course, you always retain your essence, but with the new name it becomes easier to shed some of the many superstructures we all have.

A name is not simply taken, it is conquered. The process has evolved over time and can be quite personal. Today, it usually happens like this: After a Damanhurian meditates on and chooses an animal (or plant) name based on affinity, common traits, or those desired. The feedback process subsequently begins where support from individual community members is requested for the name. This creates connection, discussion, debate, participation, laughter, memories and storytelling about names already taken. Citizens show their approval by offering something for the common good, i.e. hours of work, a piece of art, a contribution to the maintenance of the Temples of Humankind, etc.

When there is a critical mass of support, including a personal offer, the person is invited to go up on stage during one of the community assemblies, and there the name is publicly requested. Then debate opens, moderated by one of the community orators, and people from the audience can suggest alternatives. These moments are like a game and create glue amongst all the people. When the moment of decision arrives, a kind of applause meter system measuring community approval determines the wining name. If the majority prefers a different name than the one asked for, it's up to the person whether to accept the suggested name or wait for another time to ask for the one originally wanted. Every person has a choice.

Connection with the environment is expressed in many ways throughout the community. Whilst many group houses are quite old, over time they are renovated with ecological and sustainable materials. New houses are built using the latest ecological methods. Almost all homes use alternative power sources, such as solar and photovoltaic panels. Some have geothermal plants and wind and water turbines. Many harvest rainwater and the new houses have dual water systems: drinking water for the kitchen and washing, and rainwater for the WC and irrigation. When possible, groups take the water from springs and wells, and there is also a reedbed sewage treatment system serving for several houses in one region.

About 50% of the food consumed is grown inside of Damanhur with organic agriculture and many groups have their own gardens and greenhouses. Additional items are purchased in the communities' organic market, which is open to the whole valley.

Research

Born inside of 'The School of Meditation', the spiritual pathway of Damanhur, there are groups of people who experience personal and spiritual growth through a specific field of research. These paths are called the 'Ways'. They allow each person to develop talents that can be put into service for the betterment of the individual and the entire community. Every Damanhurian chooses a 'Way', some even two.

> The 'Way of the Oracle' is closely connected to ritual aspects, to the relationship with forces of the spiritual ecosystem.

> The 'Way of the Knights' is intimately involved in building and maintaining the 'Temples of Humankind' and in the protection of Damanhurian territories.

> The 'Way of the Monks and Esoteric Couples' is involved with the research of vital energy, individually or in a couple. This energy can be transformed and used for growth.

> The 'Way of Health' includes people involved in wide-ranging research that covers all aspects of healthcare: researchers, healers, doctors, holistic operators.

> The 'Way of Art and Word' is the path for those who want to research spirituality through all forms of art, including music, written and spoken word, politics, teaching, and theater.

There are also smaller groups that take the name of 'Indirizzo', which can be translated as 'branches'. These are:

> A group focuses on all areas linked to education 'Indirizzo Education';

> A group that focuses on growth in their daily work and is called 'Indirizzo Work and Art';

> A group that focuses on self-sufficiency, ecology and agriculture called

'Indirizzo Olio Caldo' ('Hot Oil', a name derived from a Damanhurian myth).

In Damanhur, spirituality and community are permanently linked together. Community is a fractal of the wholeness of being, making each action taken a manifestation of spirituality.

Macaco Tamerice (Martina Grosse Burlage) has been living in the Federation of Damanhur as a full time citizen since 1993. For several years she has been working in international relations for the Federation of Damanhur and is the Coordinator of International Community Relations. In 2008, she became Vice President of GEN Europe, and in 2010 was elected President.

Macaco is an international speaker, facilitator and Geese Educator. Trained in music and voice, she has toured as a professional jazz singer in Europe, Japan and Canada, for over twenty years. She has also taught voice and led classes and seminars since 1984. Fluent in five languages, trained in community-building and conflict resolution during the last 18 years, Macaco has held roles of social and artistic responsibility in the community of Damanhur. In 1992, she graduated from the Damanhur School for Spiritual Healers.

Auroville is an international community in South India that was founded as a centre for human unity. It was inspired by the French visionary, The Mother, and by the work and living presence of the great Indian philosopher, yogi, and revolutionary, Sri Aurobindo.

A Brief Snapshot of Auroville, India

MARTI

Auroville is a place for karma yoga, the transformation of consciousness through action. All life is seen as yoga. Through conscious and collective work, we try to bring the divine into matter. In *The Mind of the Cells*, the Mother describes how the cells in our body will carry the light and love of transformation when we have truly learned how to quiet our minds and live as one body. Sri Aurobindo describes an actual collective physical mutation of the human species, which will happen when humanity reaches a critical point of awareness. He says that this quantum leap can only happen through collective consciousness.

Auroville is a living laboratory of evolution that provides us with the opportunity to give concrete manifestation to these ideas. The Charter of Auroville, which has guided the community since its outset in 1968, says that "Auroville belongs to no one in particular, but to Humanity as a whole, but to live in Auroville one must be a willing servitor of the divine." Auroville has been described as a living laboratory of evolution and a place of unending education and a youth that never ages.

Living up to these ideals is particularly challenging, especially given that our small ecocity is a highly diverse cultural entity and even simple concepts such as home and love evoke different ideas for different people. But this is our challenge. Auroville is in many ways a microcosm of the world at large. We believe that if we can make a significant change in ourselves, the world will change, as well. We have all the elements to make that possible. We are a crossroads between East and West, drawing from ancient wisdoms and modern traditions. We are a bridge between North and South, drawing from the latest technology and located in a rural area where people have lived for thousands of years with a highly developed sense of ecological consciousness. The early work of Auroville pioneers who transformed areas of dry barren dusty red desert into lush green forests was the first sadhana.

Auroville is the largest intentional multicultural community in the world. More than three million trees were planted at Auroville over the years in its remarkable restoration project. Experiential education is highly emphasized and children have a unique opportunity to nourish their minds and spirits in community and nature.

On December 30, 2011, Cyclone Thane left a trail of devastation as it swept through Auroville. Several hundred thousand trees fell, farms and crops were severely damaged, and 150 electricity poles came down. Approximately 200 houses, workshops and public buildings were damaged, but the Matrimindir meditation sanctuary was left unscathed. Miraculously, nobody was injured, though some were trapped in their houses. The community is working hard to rebuild itself, and relief support has poured in from generous organizations and individuals around the globe.

The indomitable spirit of the Auroville community is exhibited in its rapid response to this crisis, and the sustained determination of the Auroville people to rebuild their community. This vibrant response has been an inspiration to all, and the community is thriving even as it continues its long-term restoration.

Please see page 80 for MARTI's biography.

MARTI introduces Thich Nhat Hanh's precepts for a collective spiritual life, describing Plum Village, the community he founded in southern France.

Plum Village: A Spiritual Perspective on Community

MARTI

Plum Village is a Buddhist intentional community set in a farming area in Southern France. It is also a retreat center where people from diverse traditions can come and experience the silence and daily inner discipline that leads to a deeper understanding of oneself. Plum Village is based on the Buddhist concept of sangha, or collective spiritual life. Meditation, mindfulness, and conscious respect for the environment are essential. Participants say that 'when we practice breathing, smiling, and living mindfully with our family, then they become our sangha'. In *A Joyful Path*, the Buddhist monk, Thich Nhat Hanh, founder of Plum Village, describes the five essential awarenesses:

We are aware that all generations of our ancestors and all future generations are present in us.

We are aware of the expectations that our ancestors, our children, and their children have of us.

We are aware that our joy, peace, freedom and harmony are the joy, peace, freedom, and harmony of our ancestors, our children, and their children.

We are aware that blaming and arguing never help us and only create a wider gap between us, and that only understanding, trust, and love can help us change and grow.

Thich Nhat Hanh has said, "Those who practice mindful living will inevitably transform themselves and their way of life."

A Joyful Path: Community, Transformation & Peace, Thich Nhat Hanh, Parallax Press, 1995, 978 0938077 76 3.

Please see page 80 for MARTI's biography.

Hildur Jackson first met the founder of Sarvodaya in 1996. Hildur and the Global Ecovillage Network has retained close contact with the movement ever since. Here she introduces the Sri Lankan Sarvodaya philosophy and vision. She explains how it is a whole new way of thinking and organising society and politics, which could be a new global model.

The Awakening of the Person, the Village, the Nation and the World:

The Sarvodaya Vision for our Global Future

Hildur Jackson

The Sarvodaya Shramadana Movement is the largest people's organization in Sri Lanka. Sarvodaya is Sanskrit for 'awakening of all' and Shramadana means to donate effort. It began in one village and has grown to more than 15,000. The movement dates all the way back to 1958 when Sri Lankan science teacher, Dr Ari T Ariyaratne, took his students to work camps in the poorest rural areas. The central focus of the process was the village and the village development program. By the 1970s, the Sarvodaya Movement was drawing attention of foreign scholars and donor agencies who recognized it as an example of 'appropriate' people-centered development and an alternative to the capital intensive schemes favored by the World Bank. The second generation of western aid workers have been less willing to support Sarvodaya but this has allowed the movement to independently develop their own system of governance. Today almost 16,000 villages are linked in a system for local development and their philosophy and practice are spreading all over the world.

The Goal is the Political Empowerment of the Village

For Sarvodaya development is a process of ongoing personal and social awakening of the individual, the family, the community, and the national and global societies. Since its inception Sarvodaya has tried to offer an alternative to top-down development. The Sarvodaya model of development is envisaged as encompassing six dimensions: spiritual, moral, cultural, social, economic and political.

Ari Ariyaratne's son, Vinya, explains how Sarvodaya works: "We lay three foundations in a village. The first is the psychological infrastructure in the community which consists in bringing villagers together through voluntary community action. This is essential to build the social infrastructure, which brings a bit more organization to the community. The social infrastructure is based on organizing villagers in groups (mothers, youth, farmers, elders, pre-school), where people can come and discuss their problems.

> *"Then we get the whole village together in a village society and get them legal recognition: they become a small democratic unit for self-government. We introduce basic elements of community organization: rules, election, the keeping of accounts. The third foundation, based on the psychological and social infrastructures is the economic infrastructure: we develop a banking scheme, a credit scheme and such."*[1]

The spiritual dimension runs through all these parts of the development. Villagers get together and meditate and share their spiritual practice. It's an integrated process.

Meditation, Peace Marches and Engaged Buddhism

Sarvodaya represents one of the earliest expressions of what has become known as socially engaged Buddhism. George Bond writes about the foundation for Sarvodaya thinking in *Buddhism at Work*. He describes Sarvodaya as breaking the stereotype of what Buddhism being world denying, 'all life is suffering'. Sarvodaya teaches people how to engage in the world whilst being detached from outcomes. This spiritual detachment combines with social activism is synthesis of Buddhist and Gandhian principles. From Buddhist heritage, Sarvodaya adopted the view that suffering represents the basic human predicament. Whilst from its Gandhian heritage, it adopted the view that suffering has social and structural causes that must be addressed if liberation is to be achieved. Sarvodaya's vision calls for the awakening of individuals and communities to bring about non-violent revolution. Ari called for this revolution to 'build a society whose value system is based on Truth, Non-violence, and Self-denial… a no poverty, no-affluence society'. These aims represent both an interpretation of the Gandhian ideal and an application of the Buddhist Middle Way in relation to social and economic life.

Dr Ariyaratne notes that whereas the Western economic models depends on the creation of desire, Sarvodaya's aim is to eliminate both desire and suffering. This is a radical process that awakens the individual, the community and ultimately that of humanity. How does Sarvodaya philosophy support this process?

The Personal/ Individual Awakening

Every human being has the possibility to attain supreme enlightenment. An average person, however, cannot achieve this kind of awakening in one life, the journey unfolds over many lives. Therefore Sarvodaya teaches that before people can awaken to the supreme supramundane dimension of truth they must awaken to the mundane dimensions of truth that surround them in society. To illustrate this idea, Sarvodaya has given the Four Noble Truths of Buddhist's philosophy social interpretations. People must awaken to the mundane truth that surround them in society before people can see the meaning of the Four Noble Truths.

The First Noble Truth

Dukkha (suffering or un-satisfactoriness) is translated by Sarvodaya as: 'There is a decadent village'. This concrete form of suffering becomes the focus of mundane awakening. Villagers are encouraged to recognize problems such as poverty, disease, oppression and disunity in their immediate environment.

The Second Noble Truth

Samudaya (the origin of suffering) in Sarvodaya signifies that the decadent condition of the village has one or more causes. Sarvodaya teaches that the causes lie in factors such as egoism, competition, ignorance and disunity.

The Third Noble Truth

Norodha (cessation). In traditional Buddhism an indicator of Nirvana is the cessation of suffering. Sarvodaya acknowledges that the villagers' suffering can cease and suggests that the pathway to ending suffering lies in social activism.

The Fourth Noble Truth

The Noble Eightfold Path. Joanna Macy offers an excellent example of mundane explication of the Noble Eightfold Path when she cites a Sarvodaya teacher's explanation of sati (right mindfulness). "Right Mindfulness... means stay open and alert to the needs of the village... Look to see what is needed – latrines, water, roads..."

Inner and Outer Liberation at the Same Time

If people can awaken to the mundane truths about the conditions around them and realize the need for change, they can work in society for spiritual and social liberation. As society is changed, the individual is changed. One who addresses mundane problems with compassion finds the mundane world becoming more compassionate. In a more compassionate world, it is easier for the individual to develop wisdom. Dr Ariyaratne explained the interconnectedness of dual liberation when he said, "The struggle for external liberation is a struggle for inner liberation from greed, hatred and ignorance at the same time." Thus the outer manifestation of harmony through social activism and ecovillage living nurtures the inner spiritual transformation of the individuals in the community. From this perspective, ecovillage living becomes a pathway to spiritual development.

As society is changed, the individual is changed. One who addresses mundane problems with compassion finds the mundane world becoming more compassionate.

The Awakening of the Village

How does a village wake up by becoming a member of Sarvodaya? As early as 1967, Sarvodaya launched its Hundred Villages Development Scheme to mark the birth centenary of Mahatma Gandhi. Dr Ariyaratne selected 100 villages across the country and organized workcamps (*shramadanas*) and village awakening (*gramodaya*). This served as a laboratory for refining the village development techniques, which Sarvodaya now employs in 16,000 villages. This constitutes one third of all villages on Sri Lanka.

Physical work on village projects is combined with gatherings of the families in the village. These meetings help the villagers overcome their sense of helplessness by realizing their vast potential for self-development based on self-reliance, mutual cooperation, and the harnessing of local resources. Following a workcamp, Sarvodaya prepares the social infrastructure for village development by organizing various groups. They typically include groups for preschool children, school-age children, youths, mothers, farmers and elders. These groups provide peer communities that facilitate the awakening of members and they serve to establish some of Sarvodaya's basic services for the village. Here villagers can come and discuss their problems. Everyone is involved and the fact that the youth group is motivated to work for improvement of the village rather than being disaffected is important. The involvement of the preschool and mothers groups is also important as this has empowered mothers to take a leadership role in the movement. Today, many of Sarvodaya's groups are led by women. All of these groups are subsequently organized into a village society that gets legal recognition.

At the next level of development, Sarvodaya introduces basic elements of community governance such as simple rules, an election process, and the keeping of accounts. The third level is the introduction of a local banking and credit scheme: the goal is village development rather than driving the village to enter the global market.

Ecumenical Spirituality

In an interview in *Ecovillage Living*, Vinya Ariyaratne describes the far reaching role of spiritual practice in the movement. "The spiritual dimension runs through all these parts of development. People get together and meditate and follow this spiritual practice all the way through. A camp can last for three days or a week. We start the camp with meditation combined with traditional cultural activities and indigenous practices. In a Hindu community we bring out the Hindu philosophy to awaken the inner personality. We train people from all ethnic groups in this work, so our entry point is development, not a particular religion, and spiritual development is brought in as part of village development. Every day a couple of hours are spent on spiritual and cultural activities. The basic meditation technique we follow concentrates on breathing in and out (mindful breathing).

A person belonging to any religion can practice it. We meditate together and then each religion is given five minutes for their practices. Hindu chants followed by Christian prayers and so on. This promotes inter-religious harmony in our country in a big way. We never had any problem integrating spiritual development for people from different religions. The media are promoting misconceptions about religions."

Ari Ariyaratne adds, "What is most important is the essence of religion, which is spirituality. Whatever one may call it, cosmic consciousness, or universal mindfulness…" Bond explains, "The assumption of an underlying spiritual unity tempers Sarvodaya's use of Buddhist philosophy for Ari believes that the Buddhist concepts represent universal values. As he said, "Loving kindness is loving kindness. Compassion is compassion." As soon as we apply such qualifying terms as Buddhist, Hindu, Islamic, Christian etc. to these qualities they cease to have any meaning. All that happens thereby is that the spiritual label is given protection."

Awakening of the Nation and the World

With the village development work growing, in 1978 Sarvodaya had to build a huge infrastructure of administrative centers and training centers to support the village work. For example, it built a world class Development Educational Institute and Farm in Tanamalwila that has facilities for 400 trainees.

Sarvodaya discovered that monks often played a vital role in organizing development but they needed more training to become effective. Sarvodaya received funds to build a leadership training institute for Buddhist monks that opened in 1974. Other successful national programs were for orphans and destitute children, the Sarvodaya Women's movement, and the peace Brigades, all set up to help the village societies.

In 1978 Sarvodaya built its headquarter in Moratowa. This is a large complex built as an octagon representing the eight sided wheel of Dharma, a symbol of the Noble Eightfold path. The complex houses training centers, a library, a media center, conference halls, residence halls and administrative offices. A sign on the front declares, "This abode is named Damsak Mandira and is built in the shape of Dhamma Chakka (Wheel of Doctrine). It is for young men and women, who strives to establish a Sarvodaya social order in Sri Lanka and in the World in keeping with the noble eightfold path of Buddhist philosophy."

National and Global Politics

From a Sarvodaya political point of view, the recent waves of globalization and modernization have further reinforced the hierarchical structures of society by imposing a layer of economic and consumerist oppression on top of the subjugation of the people by the government hierarchy. Thus the people at the grassroots level are trapped between oppressive state forces and oppressive market forces. Sarvodaya seeks to liberate people from the control of both the state and the market. Sarvodaya wants the people to build, or rebuild, a dharmic civil society by reconstructing a horizontal axis of village or people's power. By doing this people can free themselves from these hierarchical forces and take charge of their own governance by creating an alternative economic and political system. This is not a recent idea, but part of its vision of the awakening process articulated early on.

Peace Meditations and a New Peace Center

Sri Lanka has suffered from violent ethnic conflict for 26 years until 2009. As opposed to the logic of 'war for peace', Sarvodaya has initiated a new people's peace initiative in the form of peace meditations. The first peace meditation took place in 1999 gathering 200,000 people from various ethnic and religious backgrounds. This is now a regular event. During peace meditations people sit in silence and engage in a meditation for peace for an hour. During the meditation people are asked to erase from their minds all the divisive elements in their society and instead to look at humanity and nature as interconnected. Everybody extends their loving kindness to all living beings and directs spiritual energies towards the raising their spiritual consciousness, praying for the unity of all people so that an end to the war can be brought about. Peace will not come if only a symptomatic approach is taken. Gun control and the dispersion of violent groups will not work if there still is war in people's minds. Building peace in people's minds is Sarvodaya's approach to peace. In 2002 they had the biggest ever peace meditation in the world with more than 600,000 people present, most of them women. Many people borrowed money to get a bus ride to the meditation site. The war ended in May 2009.

In 2002 they had the biggest ever peace meditation in the world with more than 600,000 people present, most of them women.

The Tsunami and The Ecovillage at Lagosawatte

The Tsunami disaster struck Sri Lanka on the 24th of December 2004. Over 40,000 people were killed and over 500,000 left homeless. Sarvodaya played a leading role from relief to long term recovery by building the first post-tsunami ecovillage at Lagosawatte on the south eastern part of Sri Lanka in the Kalutara District.

Ecovillage design consultants, Max Lindegger and Lloyd Williams commented, "Lagoswatta provides Sarvodaya with an important opportunity to create a model village which demonstrates best practice design principles in its built environment, ecological, and social elements. It is an ideal site to introduce innovations in housing design which are practical and economically and environmentally responsible. The introduction of earth wall techniques such as rammed earth; the collection of potable rainwater; better treatment of wastewater; and better ventilated and cooler housing are prime candidates."

The layout is a greenfield site of eight acres containing 55 houses located about five kilometres from the main road in one of the worst hit tsunami districts. Besides these 55 energy efficient houses, the design includes a multipurpose community centre and a landscaped edible and recreational environment. The community center is used for:

- exhibitions on nature and the environment
- training in permaculture
- training of young architects
- revival of spiritual, cultural and social values
- maintaining cohesion in a diverse community.

The planning was completed with the future residents and construction was completed within 12 months.

The promising results of Lagoswatta encouraged other villages to follow suit, adopting ecovillage principles. Also it has been used as a model in post-war reconstruction, thus encouraging the construction of more ecovillages in Sri Lanka. Sarvodaya has also been active in Japan helping them rebuild after the Earthquake of 2011.

Sarvodaya offers a model of development to nations and the world as a whole. The model combines village development with spiritual awakening. Savodaya have demonstrated practically how this can achieved, how to build a social organization around this model and how to link development to a process of spiritual awakening that can be accepted by all faiths. Political parties are substituted by different levels of village representation.

Sarvodaya is determined that it must go forward and realize its vision on national and international levels. The movement hopes to transcend the government and global forces that they believe cause poverty and suffering and are leading society in the wrong direction. There are many similarities between the Savodaya Movement and the ecovillage movement

I see no difference between what they have been practising and building up for decades and the ecovillage vision. We therefore have a lot to learn from Savodaya's many years of experience.

Sarvodaya offers a model of development to nations and the world as a whole. The model combines village development with spiritual awakening.

Resources

Ari Ariyaratne, *Collected Works* I-VII
George D Bond, *Buddhism at Work, Community Development, Social Empowerment and the Sarvodaya Movement*, Kumarion Press, 2004.
Joanna Macy, *Dharma and Development: Religion as Resource in the Sarvodaya Self Help Movement*, Kumarion Press, 1991.
Hildur Jackson and Karen Svensson, interviews with Vinya and Ariyaratne in *Ecovillage Living*, Green Books, 2002.
For more about Savodaya see: www.sarvodaya.org

Please see page 46 for Hildur Jackson's biography.

A Hopi Elder Speaks

You have been telling people that this is the Eleventh Hour,
Now you must go back and tell the people this is the Hour
And there are things to be considered.

Where are you living?
What are you doing?
What are your relationships?
Are you in right relation?
Where is your water?
Know your garden.
It is time to speak your Truth.
Create your community.
Be good to each other.
And do not look outside yourself for the leader.
This could be a good time!

There is a river flowing now very fast
It is so great and swift that there are those who will be afraid.
They will try to hold on to the shore.
They will feel they are being torn apart and will suffer greatly.
Know the river has a destination.
The elders say we must let go of the shore,
push off into the middle of the river,
keep our eyes open and our heads above water.

The time of the lone wolf is over. Gather yourselves
Banish the word struggle from your attitude and your vocabulary.
All that we do now must be done in a sacred manner and in celebration.

We are the ones we've been waiting for.

attributed to an unnamed Hopi elder,
Hopi Nation, Oraibi, Arizona

More books from Permanent Publications

 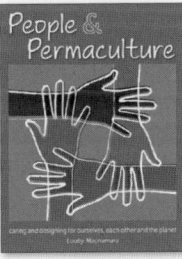   

ALL THESE & MORE AVAILABLE FROM:

www.permanentpublications.co.uk

AVAILABLE IN THE U.S.A. FROM:

www.chelseagreen.com

Practical Solutions for Self-Relinace

Permaculture is a magazine that helps you transform your home, garden and community, and save money.

Permaculture magazine offers tried and tested ways of creating flexible, low cost approaches to sustainable living, helping you to:

- Make informed ethical choices
- Grow and source organic food
- Put more into your local community
- Build energy efficiency into your home
- Find courses, contacts and opportunities
- Live in harmony with people and the planet

Permaculture magazine is published quarterly for enquiring minds and original thinkers everywhere. Each issue gives you practical, thought provoking articles written by leading experts as well as fantastic ecofriendly tips from readers!

permaculture, ecovillages, ecobuilding, organic gardening, agroforestry, sustainable agriculture, appropriate technology, downshifting, community development, human-scale economy ... and much more!

Permaculture magazine gives you access to a unique network of people and introduces you to pioneering projects in Britain and around the world. Subscribe today and start enriching your life without overburdening the planet!

Subscribe securely online at:
www.permaculture.co.uk/subscribe
or call 01730 823 311

North American subscribers contact Disticor Direct: dboswell@disticor.com

A digital subscription is available at **www.exacteditions.com** for £10.00
or **http://bit.ly/pocketmags-permaculture**

For daily updates, vist our exciting and dynamic website:
www.permaculture.co.uk